FROGS AND TOADS

FROGS AND TOADS

Look for these and other books in the
Lucent Endangered Animals and Habitats series:

FROGS AND TOADS

BY REBECCA K. O'CONNOR

Endangered Animals & Habitats

LUCENT BOOKS®

THOMSON
——✳——
™
GALE

San Diego • Detroit • New York • San Francisco • Cleveland • New Haven, Conn. • Waterville, Maine • London • Munich

LIBRARY OF CONGRESS CATALOGING-IN-PUBLICATION DATA

O'Connor, Rebecca, K.
  Frogs and toads / by Rebecca K. O'Connor.
    v. cm. — (Endangered animals & habitats)
  SUMMARY: Frog types, loss of habitat, dangers from humans, research and
conservation of frogs and toads.
  ISBN 1-56006-919-8 (lib : alk. paper)
  1. Anura—Juvenile literature. 2. Endangered species—Juvenile literature. [1. Frogs.
2. Toads. 3. Endangered species.] I. Title. II. Series.
  QL668.E2 O26 2003
  597.8—dc21

                                                  2002151096

Contents

Introduction

FROGS AND TOADS are common animal species that humans have always taken for granted. Throughout history children have captured tadpoles in ponds and puddles, bringing them home to watch their amazing transformation into frogs. The nighttime chorus of the chirps and croaks of frogs has always been a part of the soundtrack of human communities. Today, however, many places in the world where the night was once filled with the voices of frogs have gone silent. Scientists from field biologists to laboratory researchers are trying to learn why and what can be done about it.

Because many other species on the planet are suffering from the consequences of human activity such as habitat loss and pollution, many researchers initially assumed that the majority of frog disappearances had the same causes. This assumption changed in the 1980s when frogs began to disappear from habitats that had not been altered by humans. With the sudden and mysterious disappearance of the golden toad and many of its cousins from the virtually untouched cloud forest of Costa Rica, a new question was raised. If loss of habitat and pollution is not killing frogs, then what is? Scientists who study frogs began to look deeper and found an alarming trend. Frogs and toads around the world were in decline. In 1989 the alarm was raised.

Frog mutations

Shortly after the diminishing populations were labeled "amphibian decline," another disturbing trend appeared. In midwestern states such as Minnesota a large population of frogs with strange deformities was discovered. Frogs had extra hind legs or no hind legs at all. Some frogs were found with extra eyes and odd-looking growths on their bodies. The media spread the news of this amphibian "plague" and much of the country learned about the "mutant frogs." People began to look for mutated frogs in their own backyards and found them there. Twenty-six states reported the mutations and the public demand for answers grew. What is wrong with the frogs? people asked, and more importantly, Is there something wrong with the water?

A deadly fungus

In 1997 another surprising event occurred in Panama at an amphibian research site called Fortuna. At Fortuna researchers watched as 85 percent of the frogs in the study died. Laboratory examination of the dead frogs revealed that frogs were

Male golden toads compete with each other during breeding season. The sudden disappearance of the amphibian from Costa Rica prompted researchers to closely examine the cause.

This mutant frog has only one hind leg. Frogs with extra eyes, missing limbs, or other abnormalities were found in twenty-six states in the 1990s.

facing another problem, a fungus that killed them as soon as they reached adulthood.

People began to worry about the possibility that new and devastating frog diseases could cross over and infect humans, but scientists tried to ease their fears. What was affecting the frogs was most likely a combination of environmental factors. Frog diseases were unlikely to affect humans, but scientists agreed that this sudden decline of frogs might not be a natural occurrence and might instead indicate bigger problems in the environment that could ultimately affect other species. Research was crucial.

Research for the future

Today a variety of research is in progress. Though more has been discovered about the seriousness of frog and toad decline than about ways to stop it, some scientists feel that within five years we will be able to identify and explain the causes of the declines. Whether or not science will be able to reverse frog and toad decline remains to be seen. Only further research will reveal the possibilities for preserving endangered frogs and toads.

1

Frogs and Toads: The Order of Anurans

FROGS AND TOADS are ancient and diverse species known throughout history to exist just about anywhere there is freshwater. More than forty-eight hundred different species of frogs and toads have been identified and are listed on a taxonomic register compiled by the American Museum of Natural History and more are discovered every year. Approximately three hundred of these species are found in the United States. Until the recognition of amphibian decline in recent decades, herpetologists, scientists who study amphibians and reptiles, focused their research primarily on the many unique and specialized adaptations that have enabled frogs and toads to survive and thrive so well for so long.

Scientific classification

Frogs and toads belong to the family Amphibia and share the dual life cycle common to all amphibians. The word *amphibian* comes from Greek words meaning "both" and "life." The name comes from the fact that most amphibians live their adult lives breathing air through lungs and even living on land, but begin their life cycle as aquatic larvae breathing with gills before undergoing metamorphosis, meaning their body changes into a very different form to become an adult. Amphibians thus lead a double life, depending on water for part of their existence and air for the rest of their existence. However, even on land adult amphibians must have access to water or moisture to survive and to lay their eggs.

Amphibians, which are vertebrates, having backbones, share several characteristics that distinguish them from other classes of animals. For example, unlike fish, amphibians have glandular skin without scales. Amphibians are all cold-blooded, meaning they cannot generate their own body heat and must depend on the temperature of their surroundings to warm up or cool down. Unlike reptiles they also have gills during some part of their life and produce eggs without a shell.

Frogs and toads belong to the order Anura and possess characteristics that separate them from other orders within the class Amphibia. The name Anura means "without a tail." Anurans do not possess a tail in their adult stage, as do other amphibians. They also have fewer vertebrae than other amphibians. The most obvious difference between frogs and toads and other amphibians is anurans' ability to jump. Anurans have elongated anklebones, part of specialized hind legs, which are utilized to propel the animal into a jump.

A frog leaps into the water. Part of the Amphibia family, frogs and toads must have access to wet environments in order to survive.

The California Red-Legged Frog: An Example of an Endangered "True Frog"

The red-legged frog, *Rana aurora,* is the largest native ranid in California. It has the body shape and coloration that people normally imagine when they think of frogs, with long legs, smooth skin, and mottled coloration. This frog is thought to have been the species Mark Twain made famous in his story "The Celebrated Jumping Frog of Calaveras County." It is also considered one of the best frogs for human consumption and was once killed for food.

The red-legged frog has declined from 75 percent of its range in California. Scientists believe that the frogs are suffering from habitat loss, habitat fragmentation, introduction of exotic predators, and overexploitation. Currently there are only three known populations in California with adults numbering more than 350 individuals. The U.S. Fish and Wildlife Service lists the species as threatened.

Populations of the red-legged frog have declined due to habitat loss and overexploitation.

Anurans share the class Amphibia with two other much smaller orders, Gymphiona (wormlike burrowing animals) and Caudata (salamanders and newts). These other orders of amphibians can be found in many of the same moist habitats that frogs and toads are found. However, most of the more than fifty-five hundred species of amphibians are frogs and toads.

Is it a frog or a toad?

There is no scientific distinction between toads and frogs. Though anurans look very different from other amphibians, frogs and toads can look very similar. Experts commonly refer to the group as a whole as "frogs."

What people normally think of as "typical frogs," such as the California red-legged frog, are from the family Ranidae. Frogs from the family Ranidae have smooth moist skin and long, strong, webbed hind legs. Typical frogs live in moist climates, lay their eggs in clusters, and have teeth.

The Natterjack Toad: An Example of an Endangered "True Toad"

The natterjack toad, *Bufo calamita,* is a good example of the family Bufonidae, the true toads. The natterjack has a flattened body shape with somewhat dry and bumpy skin as well as a yellow stripe down the center of its back. The toads measure about two and one-half to three inches. Like most toads, the natterjack lives most of its life on land, in holes beneath roots or hidden in the grasses, but returns to the pond to breed.

This species is also the rarest amphibian species native to the United Kingdom and is considered endangered. The toad depends on sand dunes and salt marsh habitats and has declined along with the disappearance of these habitats. The natterjack also faces pollution, predation, and being run over while crossing the roadways when they migrate en masse to the breeding ponds. The natterjack is the only amphibian species found in Ireland and is protected there as well as in Scotland and England.

The typical toads, such as the natterjack toad, are from the family Bufonidae. Toads from the family Bufonidae have dry warty skin and shorter hind legs, tending to walk more than jump. Toads live in drier climates, lay their eggs in long chains, and do not have teeth.

Other families from the order Anura are not so easily classified as frogs or toads. There are twenty-nine families of anurans and many have a mixture of the characteristics that are commonly used to distinguish a frog from a toad. Scientist Chris Mattison suggests, "It may be less confusing if we think of all tailless amphibia as frogs, and use the word toad in the narrower sense, i.e., for members of the family Bufonidae."[1] Whether they are called frogs or toads, anurans have similar characteristics and common ancestors.

Ancient animals

Frogs evolved from ancient amphibians millions of years ago. The earliest known "protofrog" is *Triadobatrachus massinoti,* a species discovered in Madagascar and thought to be about 250 million years old. Scientists refer to this species as a "protofrog" because it does not have all of the characteristics of a frog. For example, *Triadobatrachus* has more vertebrae than a frog and a short tail. The bones of the forearm are not fused together as they are in modern frogs and its back legs are not as elongated as they are in modern frogs. The species does however, resemble a frog. It is thought that *Triadobatrachus* may be the link between anurans and salamanders.

The earliest record of what scientists think of as a true frog is from the Early Jurassic period, about 188–213 million years ago. That makes frogs older than most of the major groups of dinosaurs. This ancestor has been identified from impressions of a single animal that was found in 1961 in Argentina called *Vieraella herbsti*. It is 1.3 inches in length and its anatomy is very similar to that of today's frog.

Anatomy

Anuran anatomy has evolved to enable anurans to be successful predators and survive in almost every habitat. These adaptations involve not only an anuran's skeleton but also its senses and respiratory system.

Anurans can be found in a variety of shapes and sizes. Though most frogs range in length between .78–3.15 inches (20–80 mm), the smallest frog is probably *Psyllophryne didactyla* from Brazil, which is only about .39 inches (9.8 mm) as an adult. The largest frog is most likely *Conraua goliath,* the goliath frog from West Africa, which may exceed 11.8 inches

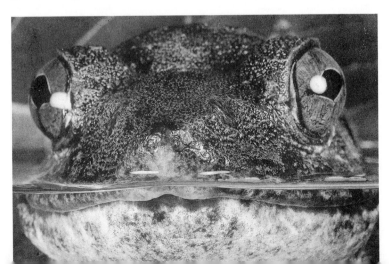

True to its name, the goliath frog from West Africa is the largest of the species. The goliath frog may measure more than 11.8 inches in length.

(300 mm) in length. Size is dependent on habitat. For example, tree frogs must be smaller on average than terrestrial, or ground-living, frogs, because tree frogs must be able to climb and support themselves on smooth surfaces and thin branches.

The appearance of frogs and toads is dictated by their environment as well. A frog's shape and color have evolved as evolutionary adaptations that enable it to compete successfully with other organisms in the same environment. For example, many frog species' coloration blends in with their woodland or desert or jungle surroundings, allowing them to evade detection by predators as well as remain unnoticed by their prey. Other frogs that do not blend in with their environments have evolved other adaptations, for example, long and powerful legs that help them escape predation or leap to catch their prey.

Limbs are a critical anuran adaptation for survival both in hunting and in escape. All frogs have two front legs with four toes each and two hind legs with five toes each, but there are many variations of this basic form. Frogs that swim, for example, have webbed toes. Tree frogs, which climb, have large round toe pads that help them move through foliage. The toe pads work like suction cups, helping the frogs cling to wet leaves. A particularly interesting limb adaptation belongs to the Costa Rican flying tree frog, which soars from branch to branch with the help of webbing between its toes that puffs up with air like a parachute, allowing the frog to stay aloft as it leaps. An anuran's body design is crucial to scientific understanding of how a species survives in the wild.

Anuran senses

Like most animals, anurans depend on their senses as well as anatomical advantages for survival. Most important to anurans is an excellent sense of hearing. With the exception of a few species, anurans have better hearing than species in the other orders of the family Amphibia. Anuran ears are "tuned" to best hear the calls of individuals of their own species. This species-specific hearing helps frogs attract and locate mates and recognize other frogs' territory.

The astonishingly varied sounds frogs and toads make—from delicate trills and chirps to staccato barks to loud, low croaks and honks famously described as "ribbet"—are also species-specific. Thus vocalization identification is one of the best ways herpetologists can tell one anuran species from another. Frogs vocalize primarily to attract mates, but also to signal aggression or warning and during ordinary encounters. Most male frogs have a sort of elastic pouch called a vocal sac connected to the mouth cavity that somehow amplifies the sound, but the mechanism is not well understood and some anuran species vocalize without a vocal sac.

Anurans also must be able to see well in order to capture the fast-moving prey that is the basis of their diet. They have outstanding depth perception, which allows them to determine how close or how far away an object is. They also have a wide range of vision, most reaching almost 180 degrees, allowing them excellent peripheral views at all times. Anuran vision is especially keen for objects that are in motion, which increases their ability to catch flying insects.

An anuran readies to catch a cricket. Anurans' excellent vision and depth perception help them to hunt prey.

Like humans, frogs are able to taste their food with their tongues. Receptor cells on the tongue, similar to taste buds on humans, translate the composition of food to a taste. Frogs are able to distinguish between sweet, sour, bitter, and salty tastes. This ability is important, alerting a frog instantly if it has captured something poisonous and increasing its ability to reject it before harm is done. Together with the frog's ability to smell, taste is also vital for nutrition, enabling frogs to taste needed nutrients such as salty minerals.

Hunting

Frogs and toads are known as very effective predators in a wide range of native habitats. Aquatic species of frogs are fast swimmers and catch prey in the water. Anurans that live on the land are able to leap after prey, use their camouflage to surprise prey, or use a long sticky tongue to grab prey.

Anurans have a variety of methods of hunting prey. During the day, most species remain hidden below the ground or under rocks, conveniently poised to snap up prey that wanders by. At night, nocturnal species may wander the damp areas in their home territory in search of small invertebrates. At other times frogs congregate in places where food sources are concentrated, such as streetlights, where toads are often seen gathering to catch insects attracted by the light. Other species ambush their prey, disguised by their camouflage even out in the open. Species such as the South American horned frog, *Ceratophrys calcarata,* lie in wait for their prey, buried in leaf litter.

Anurans prey on a wide variety of species, but because they cannot chew their diet is limited to what they can fit in their mouth. Aquatic species of frogs prey on small fish and aquatic insects. Species that live on land prey on small invertebrate species like ants, termites, and beetles. Larger species, like the ornate horned frog in Argentina, can gulp down an entire mouse in one mouthful.

A spotted horned frog gulps down a mouse in one mouthful. Frogs eat any appropriately sized animal with which they come into contact.

Frogs have such a strong feeding reflex that they will eat any creature of suitable size that crosses their path, sometimes swallowing food that is distasteful and even poisonous. In this case the frog is able to completely regurgitate its stomach. The stomach protrudes through the mouth slightly to the right and the frog wipes its stomach clean with its right front leg, then swallows the emptied organ. After this reflex happens several times, the frog learns to reject the food that triggers it on sight. Frogs' nondiscriminating appetite is actually a valuable trait, because a frog must be well fed in order to begin the reproductive cycle.

Anuran reproduction

The variety of ways frogs reproduce and develop is another example of how diverse adaptations make anurans successful in many habitats. Most frogs mate in ponds and lay eggs in water. However, some species lay eggs on land or in leaves and branches. Anuran methods of fertilization, degree of care of eggs, and breeding seasons also vary widely.

The breeding seasons of frogs fit three general categories. Some frogs are cyclical breeders and spawn, or gather to breed, simultaneously at about the same time every year because conditions that promote breeding regularly occur at that time of year. Most "true frogs" from the family Ranidae follow this cycle. Some frogs are opportunistic breeders and simultaneously spawn whenever suitable conditions occur, such as a rainy season that begins and ends unpredictably. The golden toad, *Bufo periglenes,* in Costa Rica is an example of an opportunistic breeder. Other frogs live in environments that are suitable for breeding at all times and thus are able to breed throughout the year. This is the pattern of many rain-forest species.

The anuran reproductive cycle begins with mating, a process called amplexus. The mechanics of amplexus vary slightly between the species, but always involve the male frog, which is usually much smaller than the female, jumping on the female's back and holding on there until the female lays her eggs, hours or days or possibly even weeks later, depending on the species. Some males grasp the female under the chin, others behind the legs. The males of the larger anurans, which cannot reach around the female, excrete a sticky substance that allows them to attach themselves to the back of the female. Males that have to hold on to the female with their front legs often have "nuptial pads," areas of rough skin on the toes of their front feet allowing for better grip. Once the female lays her eggs, the male releases sperm to fertilize them.

The fertilized eggs are vulnerable to predation. Usually they are laid in the open, where they represent a food source prized by fish and other animals. Frog eggs do not have a shell to protect them and are vunerable to microscopic predators as well. To survive such a high level of predation, most frogs in the family Ranidae lay thousands of eggs at a time. The more eggs the frog lays, the more likely a few of the eggs will survive to develop into tadpoles.

Anurans from other families have evolved a number of adaptations to protect their eggs. Frog eggs have a jelly coating that protects them to some degree from mold and bacteria. In some species this jelly is whipped into a foam nest. This foam makes it more difficult for predators to reach the eggs to eat them. Laying eggs in more hidden places such as on land or within foliage increases egg survival as well. Some particularly interesting species protect their eggs in even more unusual places. The female gastric brooding frog, for example, swallows her eggs, letting them develop in her stomach, and the female Surinam toad carries her eggs embedded in the skin on her back. Some species abandon their eggs where they are laid; the females of other species watch over their eggs until they hatch.

A double life: tadpole to adult

Anuran eggs hatch into fishlike tadpoles, a larval stage that is adapted to survive and develop in an aquatic habitat. All tadpoles have gills, similar to the gills of a fish, and are therefore able to breathe in the water. They are also able to swim, assisting them in escaping predators. However, unlike adult frogs, they are unable to breathe or move on land.

The frog larvae must have nourishment to develop into adults; their diet varies depending on their habitat. Tadpoles

Foam Nests: Anuran Adaptation to Protect Eggs Without Shells

There are at least five families of anurans that create foam nests and all have evolved independently. Foam nests are built in water, trees, or on land. The frothy mass that makes up the nest is similar in consistency and color to whipped egg white. The foam is formed when either the male or female frog whips secretions that are produced by the female when she lays her eggs. Most frogs engineer these nests at least near the water, so that their tadpoles have water in which to develop. However tadpoles of a few species develop in the foam nest and emerge as frogs. The foam nest is an adaptation that allows the eggs a better survival rate.

Foam nests can fulfill multiple protective functions. Foam-covered eggs and freshly hatched tadpoles are protected from drying out by the moisture-holding foam. Within the nest the eggs and tadpoles are also protected from aquatic predators, especially in nests that are in trees. Scientists also believe that the foam may contain bactericides and fungicides, which kill any bacteria or fungi that could potentially destroy the eggs.

Tadpoles develop inside eggs protected by a foam nest. The foam nest ensures that the eggs are safe from predators, bacteria, and dry environments.

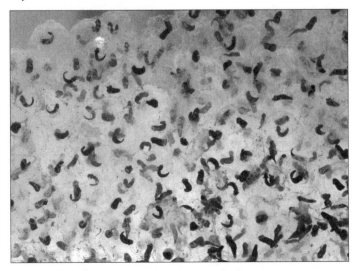

are often referred to as vegetarians, feeding on algae and plant matter, but there are exceptions. According to researcher Karin S. Hoff, "Tadpoles are better thought of as opportunistic omnivores."[2] Inspection of tadpole stomachs reveals that they eat a wide variety of plant and animal matter. Isolated in tree holes or cups of water formed by plants, some tadpoles are even fed unfertilized eggs by their mother.

Tadpoles, like adult frogs, are adapted to their specific environment. According to scientist Ronald Altig, "The morphological [body design] diversity of tadpoles is immense."[3] For example, some tadpoles have bodies designed to live in fast-moving water, like streams. These tadpoles have streamlined shapes and suckers to help them cling to rocks and keep them from being swept away. Tadpoles that live in still water have rounder body shapes and are not as powerful swimmers.

The process in which a tadpole changes into an adult frog, called metamorphosis, has forty-six distinct stages. The most

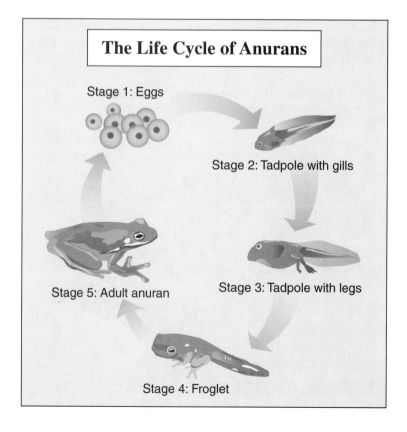

The Life Cycle of Anurans

Stage 1: Eggs

Stage 2: Tadpole with gills

Stage 3: Tadpole with legs

Stage 4: Froglet

Stage 5: Adult anuran

obvious of these stages are the development of hind legs, followed by lung development and the appearance of front legs. Gradually the tail disappears along with the gills. A froglet then emerges from the pond with just a stump of a tail, which soon disappears.

The rate at which tadpoles change into adults depends on species and conditions in the habitat in which they are developing. The change can take anywhere from a few weeks to two years. Alterations in the environment, such as a rise in water temperature, can speed metamorphosis. For example, shallow water is much warmer than deeper water; warming of the water may indicate that the pond is evaporating and shrinking. Tadpoles that normally live in temporary waters such as puddles and ponds are often triggered to rapid development by such shrinking and corresponding rise in temperature, developing lungs and within a matter of days becoming adults that can live and breathe outside of the water.

Adult frogs and toads are usually capable of breeding about a year after metamorphosis. In the wild, their life span ranges from three to sixteen years depending on species and their vulnerability to predation, but frogs have been known to survive more than twenty years in captivity.

Respiration

Since a frog's life cycle includes stages in water and on land, it is crucial that anurans are able to breathe in both types of habitat. Respiration, commonly called breathing, is the process in which an animal takes in oxygen, essential in chemical processes that produce energy, and gives off carbon dioxide, a waste product of this process. In adult frogs and toads respiration occurs via the lungs and the skin. With these two methods of respiration anurans are able to absorb oxygen both in water and on land.

The lungs in anurans are not as muscular as the lungs of mammals and work differently. Anurans do not have a diaphragm muscle or fully developed ribs and therefore cannot use the muscle and bone in their chest to expand their lungs and inhale. Instead, air is pumped in and out of the lungs when the frog raises and lowers the floor of its mouth. This can be observed in a resting frog by watching the continuous movements of its throat.

A frog's skin is an important organ involved in respiration as well. Frogs have soft wet skin allowing oxygen and carbon dioxide to pass easily between the outside air and inside capillaries, thin-walled blood cells just below the surface. The skin of a frog functions in the same manner that the soft moist lining of human lungs functions. Oxygen passes through to capillaries and carbon dioxide passes out and is expelled.

Frogs not only breathe but drink through their skin, so the condition and functioning of the skin is crucially important. One way its condition is continually refreshed is by frequent shedding, as often as every few days. The outermost layer of the skin comes away as the frog twists and bends its body to loosen the old skin. Then the frog pulls its skin over its head with its mouth, swallowing it as it comes off. Water, oxygen, and carbon dioxide are better able to pass through unblocked pores in the new layer of undamaged skin.

Anuran skin not only contributes to the transference of necessary compounds, including water and carbon dioxide, but also allows compounds introduced into the environment to pass through into the frog's body. When the waters where a frog lives are polluted, the frogs are the first to show signs of the pollution. Adult frogs may die and tadpoles may metamorphose into deformed adults. Anuran skin makes them more vulnerable to changes in the water than other species of animals.

The anuran reproductive cycle is particularly sensitive to changes to habitat. Frogs and toads are adapted to breed in very specific habitats and require reliable water sources. If water disappears due to human consumption and development or changes in weather patterns, the frogs are no longer able to breed and survive. Thus frogs and toads are one of the first species to decline if habitat is corrupted or destroyed.

2

Declining Habitat

FROGS AND TOADS have evolved to survive in every habitat except Antarctica. Anurans thrive in grasslands, wetlands, forests, and even deserts, but they are sensitive to changes to their environment. Although found in a wide variety of habitats, anurans depend on the purity of water and stable habitat in order to survive and are often the first species to be affected when habitat is lost or corrupted.

Desert habitat

Anurans are well adapted to even the inhospitable desert. *Desert* is a term that is used to describe land that receives less than ten inches of rain annually, where the evaporation rate exceeds precipitation, and which in most cases has a high average temperature. Lack of moisture in the air and ground means the earth absorbs the heat of the sun in the day, raising temperatures to levels as high as 131 degrees Fahrenheit in the shade. At night the desert floor quickly loses the absorbed heat into the atmosphere, causing temperatures to plummet, even reaching freezing. The desert is an environment of extremes, shaped by heat.

The desert landscape is also shaped, however, by water. Rains are infrequent, but often torrential. The loose and sandy soil is easily eroded in these rains. Canyons called arroyos form when water rushes down from the hills and basins called playas are formed by the rains and act as traps for water. Anurans take advantage of these desert features and infrequent rains to survive where conditions otherwise do not seem to permit survival.

A camouflaged anuran burrows into the desert sand for protection from the harsh elements. Certain species of frogs and toads are well adapted for survival in the desert.

There are two belts of deserts on Earth, one along the tropic of Capricorn in the Southern Hemisphere (including the Gobi in China, the Sahara in North Africa, and the deserts of the American Southwest) and one along the tropic of Cancer in the Northern Hemisphere (including Patagonia in Argentina, the Kalahari in southern Africa, and the Great Victoria and Great Sandy Deserts of Australia). Most of these regions are home to anuran species.

The desert seems an unlikely habitat for anurans, which require water or moisture in order to breathe and reproduce. However, frogs and toads have developed impressive adaptations, which help them to survive even in this harsh habitat. Desert anurans are mainly nocturnal, utilizing the cooler temperatures at night. These species are camouflaged to match the color of their surroundings, a useful adaptation on terrain that affords little cover. Anurans also have interesting strategies that make it possible to utilize the small amount of rainfall and dew that is available in arid climates.

According to Australian herpetologist Michael J. Tyler, "The most fundamental adaptation to periods of aridity is the capacity to burrow and, by this means, use soil as an insulator."[4] In the United States two genera of anurans that use this technique to conserve water are *Scaphiopus* and *Spea,* or the spadefoots. The spadefoots are able to quickly bury themselves and in this way they avoid the heat of the sun and conserve water. The spadefoot can remain underground for long periods when conditions are unfavorable. When the spadefoots hear the low vibrations of the summer monsoon rains, they emerge to breed in the temporary ponds. Fortunately for anurans, during this period other desert species such as termites emerge from hiding as well. According to naturalist Susan Tweit, "The toads gorge themselves on the

Named for its appendages, the spadefoot toad wades in a pool of water. Spadefoots breed in the ponds created by summer monsoons.

rich insects, eating as much as half their body weight in one night's feeding, and storing so much fat that after just a few nights of feeding, they can survive for a year or more underground."[5]

 ## When Is a Frog Considered Threatened?

A frog species is considered threatened when it meets one of several criteria. The International Union for the Conservation of Nature and Natural Resources (IUCN) publishes a list semiannually of animal species that are threatened with extinction on a global level. Most scientists recognize this list as the source for the status of frogs and other animal species worldwide. According to IUCN there are three levels in which an animal could be threatened before it becomes extinct.

The highest level of threat is "Critically Endangered." An animal is critically endangered when it is facing a high risk of extinction in the immediate future. This often means that within a ten-year period, the species has suffered or is facing a reduction of 80 percent of its population. It could also mean that there are less than 250 mature individuals of the species in the wild. The species is considered to have a 50 percent chance of becoming extinct within a ten-year period or within three generations.

The next level is "Endangered." An endangered species is facing a high risk of extinction in the near future. Generally, the species has experienced or is about to experience a reduction of 50 percent of its population in a ten-year period. Populations often number less than twenty-five hundred mature individuals. The species has a 20 percent probability of becoming extinct in the next twenty years or five generations.

Lastly, the lowest level in the threat of extinction is "Vulnerable." A vulnerable species is facing a high risk of extinction in the medium-term future. This means a population loss or predicted loss of 20 percent over a ten-year period. Also the species may have less than ten thousand mature individuals in the wild. The probability of a vulnerable species becoming extinct is at least 10 percent within one hundred years.

A frog does not have to be threatened globally in order to be given special consideration. Frogs can be considered endangered on a federal level if they are in danger of disappearing from a countrywide range. In the United States, the U.S. Fish and Wildlife Service decides what species are in danger of national extinction. There are also fish and wildlife organizations in every state that decide what animals are endangered at a state level or sometimes even just in danger of disappearing from small areas of habitat which they have historically inhabited. These organizations generally classify species as "endangered," "threatened," or "of special concern."

Some frogs take the act of burrowing one step further by forming a cocoon, which holds in as much water as possible. One species that forms a cocoon is the water-holding frog of Australia, *Cyclorana longipes.* The cocoon formation is a modification in the shedding process. Scientist R.S. Seymor found that one species of cocooned frog's "evaporative water loss was reduced by approximately seven-and-one-half times, compared to individuals lacking a cocoon."[6]

The effects of altering desert habitat

Desert-dwelling frogs are dependent on their arid habitat remaining unchanged. Places where water traditionally pools during a rainy season or irregular downpour may disappear if humans alter the terrain. If breeding grounds disappear, the species declines.

As humans develop the land, they change not only terrain, but also the climate. Golf courses, lawns, and landscaping add moisture to the environment that would not normally be there. These changes may be beneficial to some species, but can make survival difficult for frog species that are adapted to live specifically in dry desert habitats.

Many deserts, despite low humidity, high temperatures, and low rainfall, still include riparian ecosystems. Riparian means that the land surrounds waterways or streams. This riparian ecosystem of Arizona is home to at least five species of leopard frog, all of which are declining due to human development. The Chiricahua leopard frog, *Rana chiricahuensis,* has suffered the largest, most dramatic decline of all these declining species.

The decline of desert leopard frogs is mainly due to human habitation and overuse of waterways. As humans move into desert habitat and utilize existing water, this affects the small amount of running and standing water that exists in the desert. Even tapping into the underground water table ultimately affects the amount of water in running streams. Without this water there is nowhere for the aquatic leopard frogs to breed. More than 75 percent of the Chiricahua leopard frog habitat has been lost to livestock grazing, dams, and water diversions.

Despite specialized adaptations aiding in survival and the ability to utilize limited water resources, many desert anurans are declining. Some of these frogs and toads, like the spade-foot, are also able to survive in grassland habitats. However, these grassland ecosystems are declining as well.

Grasslands

Grasslands, like desert, might not appear to be attractive habitat for anurans. Temperate grasslands such as the prairies and plains of North America have an annual rainfall of only ten to thirty inches of rain, a high evaporation rate, and seasonal and annual droughts. Even the more tropical grasslands have a marked wet and dry season. Still, there are many anurans that survive in these conditions. Anurans that survive in the grasslands are adapted to survive in arid seasons.

Grasslands appear on most continents and include the prairies of North America, the pampas of South America, the plains of Europe and the steppes of Asia. Grassland species are mainly burrowing anurans that are nocturnal, or only active at night when the temperature drops.

One of the most endangered species of anurans in the United States lives in the grasslands of Texas. The Houston toad, *Bufo houstonensis,* depends on the loose sandy soils of the Texas post oak savanna. The toads must burrow in order to protect themselves from cold weather in the winter and hot, dry conditions in the summer. The Houston toad is particularly dependent on its habitat because it is a poor burrower and depends on especially loose soils that are easy to dig. Large areas of sandy soil that is deeper than forty inches are the Houston toad's preferred habitat. In recent years widespread conversion of grasslands to nonnative plant species has altered this habitat; the toads have great difficulty moving through thick vegetation or burrowing through dense roots. Due to overgrazing, fire suppression practices, and human habitation, much of the former savanna grasslands in the toad's habitat have grown into uninhabitable brush thickets. The Houston toad, found nowhere else in the world, is in danger of extinction if the grasslands of Texas continue to disappear.

Other anurans are threatened by the loss of grasslands as well. For example, *Bufo debilis,* the Western green toad, is disappearing from Cimarron National Grassland in Kansas. Human farming and ranching there has decreased the amount

The Houston Toad: Dependent on Disappearing Grasslands

The Houston toad was discovered in a very limited range of southeastern Texas in 1953. The toad is two to three and one-half inches long. General coloration varies, but usually the toads are light brown and the underbelly cream colored and mottled. Adult toads are restricted to areas of loose sandy soil and can be found in pine or oak woodland and savanna.

In the 1950s the number of the Houston toad declined sharply. A regional drought played a large role in this, followed by loss of suitable habitat due to land development. The toad also suffers from deteriorating ponds due to soil erosion, cattle grazing, and pesticide runoff. In 1970 the toad was listed by the U.S. Fish and Wildlife Service as endangered. In order to protect the toad more than two thousand acres have been purchased as a study area for the Houston toads. Scientists feel that more research must be done to better understand the toads and their life history before they can be saved.

The Houston toad is only one of many anurans that are listed on the endangered species list.

of habitat available for declining anuran species. Ranchers can harm grassland habitats by allowing cattle to overgraze. If cattle are allowed to feed in one area too long, they will eat the native grasses all the way to the roots, destroying the natural plants. Farmers have utilized the grasslands for agricultural use, also destroying grasslands. When native plants are destroyed and habitat is changed, anurans lose cover in which to hide from predators, appropriate soil in which to burrow, or natural breeding ponds. As more grasslands disappear, more frogs and toads have difficulty surviving.

Wetlands

Many species of anurans depend on wetlands for survival. Wetlands—marshes, swamps, and peat lands—are found on every continent except Antarctica and in many climates, from the warm tropics to the freezing tundra. They cover 6 percent of the land surface on the planet, or approximately 2.2 billion acres. The United States alone has 274 million acres of wetlands, about 12 percent of the wetlands on Earth.

Wetlands are a perfect environment for frogs and toads, with generally moist conditions year-round and plenty of available water. Wetlands preferred by anurans typically occur in low-lying areas that collect freshwater at the edges of larger bodies of water. In these wetlands the water table, or surface of the water, is usually at or above the land surface. This means plenty of available still water for anurans to breed in or escape into when fleeing predators such as birds and small mammals.

Since the 1700s over half the wetlands in the United States have been lost. As the habitat has declined the species that depend on the habitat have declined as well. One of the species in the United States that has diminished due to the loss of wetland habitat is the Illinois chorus frog, *Pseudacris streckeri illinoensis*. This species is currently classified as rare by the Missouri Department of Conservation and is also a candidate for federal listing by the U.S. Fish and Wildlife Service. The main threat to this frog's survival is the draining and clearing of sandy wetlands.

The Illinois chorus frog is able to burrow, but this species prefers permanent wetlands in order to breed. The frogs spend the majority of the year underground, emerging to breed in late winter and occasionally in summer rains. Historical changes in habitat resulting from channeling and filling wetlands have caused the frog's decline. The lack of appropriate wetlands has caused higher mortality, lower reproduction rate, and reduction in population size, a trend that can be seen in other wetland-dependent species as well as anurans.

Wetlands are disappearing to development in countries around the world. According to herpetologist Trevor Beebee, "Drainage of marshlands to improve crop yields and reduce diseases such as liver fluke and malaria has continued . . . to the extent that in Britain only isolated patches remain."[7] This loss of habitat has directly reduced anuran populations worldwide.

Tropical and temperate rain forests

Although the loss of wetlands concerns herpetologists, biologists are more seriously alarmed by the loss of the earth's tropical and temperate rain forests, the most rapidly diminishing habitats on the planet and home to the greatest and most diverse number of anuran species.

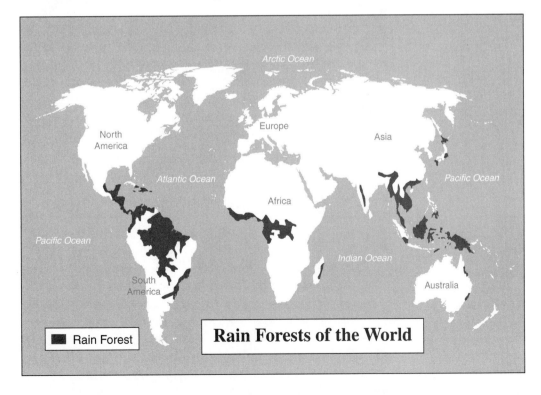

Rain Forest

Rain Forests of the World

Rain forests are normally hot and steamy, with average temperatures of about seventy-seven degrees Fahrenheit and at least five inches of rainfall per month. Vast tracts of rain forest, home to the most concentrated and varied number of animal and plant species on the planet, once existed close to the equator, where temperatures vary little throughout the year. More than any other habitat, this environment is suitable to anurans because of the availability of moisture and the constant temperature. Frogs are able to keep their skin moist and find easy access to water in which to lay their eggs. According to scientist Milan Kminiak, "More than 80 anuran species may inhabit a single hectare [2.47 acres] of rain forest."[8]

The rain forest offers anurans ecosystems that do not exist anywhere else; for example, in the overstory, or top level of the rain-forest canopy, in Central and South American rain forests, more than seventeen hundred species of plants known as bromeliads store water in small pools. This contained water

A blue dart frog clings to a bromeliad, a plant found in the rain forest. Rain forests provide the moisture and temperature levels anurans need to survive.

is both ideal breeding ground and a source of drinking water for many anuran species. When the specialized plants in the rain forest are destroyed, many species of frog diminish because they are unable to live anywhere else.

One example of rain-forest frogs that are in danger of disappearing with their habitat is the genus known as *Mantella*. About twelve known species of these brightly colored frogs frogs thrive in the rain forest on the island of Madagascar, located off the coast of eastern Africa. Since Madagascar is an island habitat, the rain forest has been isolated for millions of years and the rain-forest species have adapted to live in this very specific habitat. According to reporter Wendy Williams, "Mantellas are particularly vulnerable, since they are highly adapted to fast-disappearing specialized habitat."[9]

Rain forests around the world are rapidly being altered or destroyed to satisfy needs for food, fuel, and building materials for human populations. Land is changed and cultivated for agriculture by slash-and-burn methods, in which large areas of rain forest are cut down and then burned to the ground, destroying native animal species as well as vegetation. Unfortunately, rain-forest soils offer limited nutrients for agriculture and are quickly depleted, forcing rapid destruction of more rain forest and leaving behind damaged rain forest that cannot be restored. In Madagascar only 34 percent of the original forest cover remains.

The loss of land affects many anuran species, but it is not only loss of land habitat that contributes to their decline. No matter what ecosystem anurans inhabit, the quantity and quality of the water they depend on is crucial to their survival.

A need for water

Frogs and toads are adapted to a variety of water bodies. Certain species of frogs and toads depend on fast-moving bodies of water such as mountain streams to spawn. The tadpoles of these species have a suction mouth by which they cling to rocks so that the currents do not sweep them away. Because it takes two to three years for these tadpoles to develop into frogs, undisturbed habitat is essential to their survival. One such species, the tailed frog, *Ascaphus truei* of

California, is classified as protected by the California Fish and Game Commission because alteration of their habitat is removing the only water sources in which they can reproduce.

The larval stages of several species of Australian anurans are also adapted specifically to fast-moving waters by living in boulder-strewn areas where there is some protection from the currents. *Litoria coplandi,* for example, is found only in rocky creek beds.

Still waters are the preferred habitat of anurans. Permanent still-water bodies such as a lake or pond can support several species of anurans that do not compete with one another because such water bodies often contain a wide range of microhabitats, or different ecological niches that vary in temperature, depth, and composition, in which different species thrive. Some, such as the South American *Telmatobius atacamensis,* are completely aquatic and spend their entire lives in water. Others spend part of their adult lives on land at the fringes of the lake or pond.

Other frogs depend on temporary bodies of water, only using them for breeding, such as spadefoot toads of the genus *Scaphiopus.* The advantage of temporary bodies of water is the lack of predatory fish; many anuran species that might be easy prey do well by breeding under these specialized conditions. These frogs and toads only appear during the rainy season to breed quickly and deposit their eggs in temporary water. These eggs hatch quickly and develop from tadpole to frog in a short time, sometimes within weeks.

Altering water sources

Anuran populations diminish when water sources are altered to the point that breeding cycles are interrupted. In all anuran habitats water sources are disappearing or becoming tainted by human consumption and usage, representing the biggest threat to anuran populations.

Human consumption and diversion of water for drinking, farming, ranching, and hydroelectric plants alters anuran habitats. One species that was nearly eradicated by human consumption of water is the Kihansi spray toad, an example of an anuran that requires a very specific habitat.

Kihansi Spray Toad: A Frog Species Suffering from Loss of Habitat

The Kihansi spray toad, *Nectophrynoides asperginis,* was first discovered in 1996. It is a tiny toad, measuring only an inch long, that depends on one of the smallest habitats of all anurans. The toad is confined to an area that is just under five acres and surrounds the spray zone of the Kihansi gorge in Tanzania, where the Kihansi River creates rapids and falls.

When the river water diminished due to diversion to a power plant, the spray zone disappeared along with 95 percent of the toads' habitat. Artificial spray systems were erected as a substitute for the natural spray; however, scientists feel these cannot replace the natural falls. Believing that the best hope for the toads is captive breeding, scientists collected toads from the Kihansi site in November 2000, and 499 toads were set up in breeding colonies. Coordinated by the Wildlife Conservation Society (WCS) at the Bronx Zoo, the group was split between the Bronx Zoo and Detroit Zoo. Very little was known about this recently discovered species and researchers concentrated on not only breeding the toads, but also learning about their natural history and needs.

WCS dedicated an entire room to the toads, being extremely careful with quarantine procedures, meaning they took care not to accidentally introduce the toads to new diseases or parasites. The toads bred almost immediately and the first offspring were found on December 16, 2000. The toads give birth to live young, which is unusual for anurans and means that they reproduce in lower numbers than the frogs that lay thousands of eggs. Fifty-eight of the toads taken from the wild were still alive as of July 2002, and combined with their offspring there were 580 toads in the project, which now involves five zoos.

A Kihansi spray toad is placed next to a penny to show its tiny size.

The Kihansi spray toad, *Nectophrynoides asperginis,* lives in the spray produced by large amounts of water flow over Kihansi falls in Tanzania. According to biologist Jason B. Searle, "The spray creates an almost constant temperature and humidity, and protects the toads from predatory safari ants.

This microhabitat is so specialized that the amphibian cannot be found in any of the surrounding wetlands or gorges." [10] However, in 1999 the Tanzanian power authority began diverting water from the falls to produce electricity. When the water slowed to a trickle, the toad, which can be found nowhere else in the world, no longer had a habitat. Currently these toads are being bred in captivity and scientists are hoping that a misting system will restore their habitat.

Other water sources disappear as humans change the landscape to make it more suitable for human use. Bodies of water are drained for construction, for agriculture, and to eradicate breeding grounds for mosquitoes and thus potential spread of mosquito-borne diseases such as malaria and West Nile virus.

The problem of destruction and consumption of water sources is compounded when the quality of the water that remains is compromised by pollution. Anurans require not only sufficient water but also clean water to reproduce and develop normally, and water pollution of all kinds significantly endangers frog and toad populations around the world.

3

Threats to Survival in the Wild

ANURANS HAVE A high mortality rate in the wild. Anurans at all stages of the life cycle are a food source for many species, and despite their high rate of reproduction they naturally have a difficult time surviving in the wild. Although wild anuran populations have always faced natural pressures, scientists believe that human-induced environmental changes have made it increasingly difficult for anurans to survive in the wild, even in pristine habitats.

Natural predation

Anurans are vulnerable to a variety of predators at all stages of their lives. To deal with this high level of predation, most frog species lay huge numbers of eggs, increasing the chance that enough offspring will survive to adulthood to propagate the species. Nevertheless, predation is a constant threat to all anuran populations.

Anuran eggs are nutritious and eaten by many animals. Frog eggs that are laid in the water may fall prey to aquatic insect larvae, salamanders, and aquatic birds such as ducks. In permanent water sources, fish are the predominant predator. Where frogs use temporary pools to spawn, insect larvae, such as dragonfly larvae, are the main predators. Under extreme conditions, eggs that have been laid late in the season may even be consumed by other tadpoles.

Terrestrial eggs, or eggs that are laid under moist conditions on the ground, are also food for insect predators, especially carabid and tenebrionid beetles. Even eggs that are

A snake feeds on eggs laid in a tree by a tropical arboreal frog. Anurans are prey to a variety of animals.

wrapped in a foam nest are not completely protected from predators. In Argentina, the foam nests of the frog *Leptodactylus latinasus* are commonly eaten by a particular species of lycosid, or wolf, spider. Locusts, crabs, and crickets eat eggs laid in trees by tropical arboreal frogs. In South America annulated (cat-eyed) snakes consume the nests of leaf and grass frogs.

An even wider array of animals eat the larval anurans that manage to hatch into tadpoles. Tadpoles are the prey of many different kinds of turtles, wading birds, and small mammals. Songbirds such as the gray jay feed on the tadpoles of the boreal toad, *Bufo boreas,* and snakes such as the colubrid snake of South America are known to feed on tadpoles as well.

Many aquatic insects are also highly dependent on tadpoles for food. Tadpoles are the chief prey of diving beetles, water bugs, and water scorpions. These insects capture tadpoles, sucking the body fluids while the tadpole is within their grasp.

Perhaps the most significant predators of anuran tadpoles are other amphibians. In temporary ponds tadpoles are the most abundant food source, heavily preyed on by both adult and larval newts, which are aquatic amphibians.

Some tadpoles are themselves anuran predators. According to herpetologists William E. Duellman and Linda Trueb, "Tadpoles of a few species, such as *Ceratophrys cornuta,* seem to be obligate carnivores and feed chiefly on tadpoles of other species."[11] Even adult frogs of some species, such as the African clawed frog, *Xenopus laevis,* prey on tadpoles.

It is not surprising that a very low percentage of tadpoles survive to become adults. Duellman and Trueb state that "the survivorship of Rana larvae in ponds is generally less than 10%."[12] And predation continues to threaten adult anurans. Almost all predators eat frogs and toads, including mammals such as badgers, raccoons, opossums, hedgehogs, and bats. According to scientist Rudolf Malkmus, "Up to 100 frogs have been found stored in polecat burrows during the winter period: the frogs were still alive, but had been rendered immobile by bites."[13] More than one hundred species of European birds are known to prey on frogs and toads, based on findings of frog remains in their stomachs. Wading birds such as heron and egrets are known to prey heavily on anurans at the edges of ponds.

Certain species of snakes eat frogs almost exclusively. In the tropics, where anuran species are diverse and abundant, nocturnal tree snakes feed on arboreal frogs at night when they are active, diurnal tree snakes ferret out resting frogs during the day, and aquatic snakes prey on water-dwelling species.

Various species of predatory spiders also feed on anurans. Some spiders catch small frogs in their webs, killing and eating them. Other spiders such as crab spiders and tarantulas are active hunters, springing on their prey, grasping it, and killing by injection.

Major aquatic predators of anurans are fish, turtles, and crocodilians. Some adult anurans feed on other adult frogs, as well as tadpoles. Larger frogs are known to make a meal of smaller frogs. The hylid frogs (*Hemiphractus*) commonly consume other frogs. Duellman and Trueb note that the "stomachs of 10 specimens of *H. proboscideus* contained 15 frogs of 12 species."[14]

Survival tactics

A crocodile's jaws slowly close on a frog. The crocodile is a natural, aquatic predator of the frog.

Beyond their high reproduction rate, all anuran species have various adaptations that enhance their chances of surviving predation. For instance, anurans have developed powerful hind legs, allowing them to jump great distances quickly to escape danger.

Anurans employ several jumping strategies. Many frogs, like ranids, utilize a single long leap from land to water as a way of escape. Tree frogs will leap from branch to branch, putting distance between themselves and a predator. A further adaptation of this means is found in the gliding frogs, which use their webbed toes to glide even farther. Some frogs, like the rocket frog in Australia, cover large distances rapidly via series of short fast hops. The short multidirectional jumps of some species confuse predators, allowing escape. Many frog species jump once and then freeze, depending on their camouflage to hide them while motionless.

Many species of anurans have adaptive coloring and texture that allows them to escape notice. Frogs that live on the forest floor are often dull gray, brown, or black. In contrast, tree frogs are often green, blending in with leaves, and sand-dwelling frogs are often yellow. Some frogs' skin features intricate patterns, allowing them to conceal themselves in lichen or tree bark. Other frogs blend in to their surrounding with frills, flaps, and decorative appendages on their bodies. This extra camouflage serves to disguise the animal's outline. For example, an Asian horned toad hiding in leaf litter matches not only the color but also shape of the dead leaves in which it hides.

Many frog species distract predators by displaying a brightly colored patch of skin, called "flash colors," as they extend their legs and leap away. The predator's eyes are drawn to the bright color, but as the frog pulls in its legs, the color disappears, leaving the predator confused and unsure where the frog landed.

Many anurans have developed toxins on their skin that either taste bad or poison predators that attempt to eat them. Frequently these species are brightly colored red, orange, and yellow, advertising rather than camouflaging their identity so that predators will learn to recognize them as poisonous and leave them alone.

Perhaps the most famous of the poison frogs are the poison dart and arrow frogs from the genera *Dendrobates* and *Phyllobates*. The toxins are so strong on the skin of these frogs that the people native to Central and South America once used the frogs to poison their darts and arrows for hunting. *Phyllobates* can be particularly poisonous. These frogs produce toxins called batrachotoxins, which are among the most toxic, naturally occurring, nonprotein toxins. Ingestion of these toxins produces erratic heartbeats, rapid heartbeats, and ultimately heart failure. According to Duellman and Trueb, an individual of the largest species of poison arrow frogs, *Phyllobates terribili,* "has enough toxin to kill about twenty-thousand 20-gram white mice, extrapolated to be sufficient to kill several adult humans."[15] Poison arrow and dart frogs are active in the day and seem to show little fear of predation.

Many toads, frogs of the genus *Bufo,* excrete a terrible-tasting milky substance when they are threatened. This secretion normally comes from the paratoid glands, which are located behind the eyes. Some toads can even spray this secretion. For example, the marine toad, *Bufo marinus,* can spray its gland secretions into the face of a predator from more than a foot away. Predators quickly learn to leave this toad, which is not brightly colored, alone.

Despite all of their adaptations to survive predation, frogs and toads must continue to avoid predators on a daily basis. Although this is the natural course of anuran life, added pressures caused from changes made by humans can have disastrous effects on a population already in a daily struggle for survival. For example, when predatory species are introduced to habitats, frogs are often not equipped to evade the nonnative predator.

Introduced species

Nonnative species that have been introduced to new habitats often have a negative effect on anuran populations. Unable to evade the new predator, entire populations of frogs may disappear.

Fish are probably the most dangerous form of introduced predator to anurans as an order. Many human communities stock their streams with fish for sport fishing, for example. One of the most popular stock species is rainbow trout. Introduced trout will eat frog eggs, completely annihilating populations of stream-dwelling frogs. In Australia, where rainbow trout is not native, such introductions have had devastating consequences on frog populations. Black bullheads, another fish stocked for sport, are thought to play a large part in the decline of the Chiricahua leopard frog in Arizona.

In order to control mosquitoes, humans have introduced minnows of the genus *Gambusia* to local water bodies, with serious consequences for anuran species because the minnows devour amphibian eggs and tadpoles. In Europe, particularly in France, the introduction of goldfish to ponds has created problems for frog species. According to French biologist Alain Dubois, "People bring them home from holiday and don't know what to do with them, so they dump them into the nearest pond. Zap! The frogs are gone."[16] Introduced goldfish are also believed to carry a virus that is killing frogs and other aquatic species in the United Kingdom.

Fish are not the only new predators that are endangering anurans; introduced ground mammals are a problem as well. The introduction of rats in New Zealand is thought to have caused the extinction of populations of several species of the genus *Leiopelma*. The introduction of nonnative frogs also puts tremendous pressure on native anuran species. In the United States, the bullfrog, *Rana catesbeiana,* is thought to

Located behind the eyes, the paratoid glands of the Bufo *excrete an awful-tasting substance when predators threaten them.*

The Cane Toad: An Introduced Frog Devouring Native Anurans

The introduction of *Bufo marinus* to Australia, where it is called the cane toad, has caused a great deal of controversy. Over fifty years ago this large toad was introduced to Queensland to control insect pests on the sugarcane fields. The toad is normally native to northern parts of South America, its range extending through Central America all the way to the southern part of Texas in the United States. Its range covers a variety of habitats from tropical to arid, which implies this is a very adaptable species. It is the cane toad's adaptability that has caused Australians so much trouble.

One hundred one toads were released in Queensland in 1936. By 1974 the toads had spread to cover 33 percent of Queensland, about 226,000 square miles. Today they continue to spread at a rapid rate. The toads did not confine themselves to the sugar fields and it is arguable whether or not they had any impact on the insects affecting the fields. The toads are highly poisonous if the toxins they exude are consumed. It is possible that many native animals have died attempting to make a meal out of a cane toad. These toads are also voracious eaters and it is possible that they are eating many native species themselves, disrupting the natural balance.

Today the Australian government is seeking ways in which they might control the cane toad. Scientists are considering introducing specific disease viruses that may reduce the incredibly large populations of cane toads now inhabiting Australia.

The Australian government is attempting to control the cane toad population whose voracious appetite disrupts the ecological balance.

be a major reason for the decline of the California red-legged frog, *Rana aurora,* and foothill yellow-legged frogs, *Rana boyii.* According to reporter Ashley Mattoon, "The bullfrog is not particular about its habitat and it has a voracious appetite. It will try to swallow almost anything it can fit in its mouth. After it was introduced to California in the early 1900s several populations . . . vanished. Perhaps the bullfrog out-competed them for prey; perhaps it swallowed them."[17]

The most documented case of anuran introduction and subsequent decline is that of the introduction of the marine toad, *Bufo marinus,* to Australia. The toads were introduced to the Australian sugarcane fields in order to control a species of beetle that was damaging sugarcane crops. Although this same program worked well in Puerto Rico, it proved disastrous in Australia, where the fast-breeding toad remains out of control. Fearing that the toad will destroy existing Australian anuran populations, scientists are searching for ways to control the numbers of what the Australians call the cane toad.

Even with the best intentions, frog introduction can still be harmful. In Australia well-meaning amateur conservationists are attempting to help frog populations by introducing them back into the wild. However, they are often being released into areas in which the species was not previously found. Although the frogs are native to the country, introduction to other endangered frogs' habitats can have severe consequences.

Breed and release projects are very carefully monitored and engineered by scientists. They have to take special care that the release site is appropriate and that the tadpoles that are being introduced do not carry any diseases that can be passed on to other frogs in the area. Amateur conservationists without a laboratory and the appropriate training have the best interests of the frogs in mind, but are unable to reproduce the work of trained herpetologists. According to Gerry Marantelli, manager of the Amphibian Research Center in Melbourne, "There are many risks [to amateur releases]: gene flow into new areas, the introduction of new species, and obviously the spread of diseases."[18]

Changing weather patterns

Anuran populations naturally wax and wane depending on the weather. Scientists agree that natural cycles of drought and rain affect frog populations, which depend on heavy rains to breed. Some years frogs are abundant; some years they do not appear at all. However, several species seem to have completely disappeared during atypical El Niño conditions and scientists believe that the weather may have played a role in their disappearance.

"El Niño" is a weather pattern recently occurring at five-year intervals caused by unusually warm waters off the Pacific Coast. The warm currents change the normal climatic patterns in parts of North and South America. In the Monteverde cloud forest of Costa Rica, this weather phenomenon caused particularly dry conditions in 1986–1987.

Some anuran species have an extremely difficult time adapting to changing climate and this seems to be true of the golden toad and the harlequin frog of Costa Rica. The golden toads disappeared during the unusually dry El Niño years. Herpetologists Marty Crump and Alan Pounds studied the weather patterns between 1970 and 1990 and showed that the only cycle with abnormally low rainfalls occurred between July 1986 and June 1987.

Marty Crump studied the golden toad in 1987 when researchers counted nearly fifteen hundred in the field, but in 1988 when she returned to continue her studies she was only able to find one male frog. Since 1988 no other golden toads have been found. The harlequin frog disappeared as well. Although the evidence is not conclusive, Crump believes that the weather played a significant role in the disappearance of these two species. It is possible that the harsh conditions may have stressed the frogs, leading them to succumb to other problems such as parasites, fungi, or bacterial infections.

These two species were not the only frogs negatively affected. According to Crump, "Half of the fifty species of frogs and toads known from the area disappeared in the late 1980's, likely from the prolonged drought. Of the twenty-five species that disappeared, only five reappeared during 1991–1996."[19]

Scientists are unsure if the weather changes and their effects on amphibian populations are normal. Some scientists claim that since amphibians have survived millions of years of climate changes a few years of drought should not have had such a serious negative effect on anuran populations. Some

 The Mountain Chicken: A Species Diminishing

The mountain chicken, *Leptodactylus fallax,* is one of the world's largest frogs, measuring as long as eight inches, and is also one of the most endangered. The mountain chicken is found only on two islands in the eastern Caribbean, Dominica and Montserrat. Previously found on a number of other islands, this species has disappeared everywhere except these two places.

In 1995 a surprise eruption of a volcano on Montserrat destroyed a portion of the mountain chicken's habitat and disrupted the normal life cycle of most of the population. In 1997 the Montserrat Forestry Department began a monitoring program to track the health of the population of frogs. Initially the frogs reacted badly to the effects of the eruption. Their numbers fell and many were found burned by the acidic ash. Numbers are now beginning to recover, but the frog is now indirectly threatened by a feral pig population, released when humans evacuated. The rooting behavior of the pigs disrupts the mountain chicken's breeding cycle. Female frogs lay their eggs in a burrow and return to feed the tadpoles with unfertilized eggs. If anything happens to the burrow or the tadpoles' mother, they will not survive.

In 1999 the Durrell Wildlife Conservation Trust at the Jersey Zoo in the British Channel Islands received nine adult frogs on loan from the Montserrat government. Their studies have revealed much about this frog's natural history (such as the surprising care the female frog gives her tadpoles). Starting in June 2000 the program began to have success breeding the mountain chickens. If this first generation offspring is able to breed as well, then the research at the Jersey Zoo will be used by the French government to begin a reestablishment program.

The Golden Toad: Extinct?

The golden toad, *Bufo periglenes,* is currently believed by scientists to be extinct. Once the toad occupied a small area of cloud forest in northern Costa Rica, currently known as the Monteverde Cloud Forest Preserve. However, this toad has not been seen since 1988 and there are none in captivity.

There is very little known about the natural history of the golden toad. Since the toad was highly visible only during the breeding season, researchers were unable to learn much other than the breeding habits of this species. The toad is an extreme example of sexual dimorphism, meaning that males and females look different. The adult females are 1.8 to 2.2 inches long, black with scarlet blotches edged in yellow. The adult males are 1.5 to 1.9 inches long with a very striking orange color, which gives the species its common name. During the breeding season the males outnumbered the females eight to one.

Scientists are still not certain what caused the golden toad's disappearance. It is widely believed that the El Niño weather phenomenon was a factor in the species' decline. The causes of the toads apparent extinction are still under debate, but many scientists believe that the golden toad is an example of overall amphibian decline.

scientists suggest that climate changes are an example of the effects of global warming, the idea that humans are releasing so much carbon dioxide into the atmosphere from burning fossil fuels that Earth's temperature is slowly rising, too quickly, however, for anurans to adapt to the weather changes. According to researcher Kathryn Phillips, "some scientists speculate that amphibian declines and disappearances in places like Monte Verde just might be pieces of biological evidence that global climate change is already having serious effects."[20]

4

Threats to Survival Posed by Humans

THROUGHOUT RECORDED HISTORY frogs and toads have been affected by human admirers who collect them, eat them, or have discovered other benefits of harvesting them. Frogs and toads were once so common that these activities posed no great threat to their survival, but where habitat loss has reduced frog populations to dangerously low levels, traditional human uses of frogs have become serious threats in themselves. Until recently largely unconcerned about the plight of anuran species, however, humans have focused conservation efforts on many more high-profile species such as large mammals, and the use of frogs and toads by humans continues to threaten anuran survival.

Negative attitudes toward anurans

Despite the fact that anurans are beneficial species and cannot intentionally harm humans, most human cultures have historically held mixed feelings about frogs and toads. Paradoxically, both negative and positive attitudes have been harmful to anuran populations.

Biblical references to frogs as unsavory creatures contributed to ill feelings toward frogs and toads a millennia ago. In the Bible frogs are named as one of the ten plagues brought down upon the pharaoh in the book of Exodus. Frogs were considered impure by the ancient Israelites, who were forbidden to eat frogs according to Scripture.

A French manuscript depicts "The Plague of Frogs," one of the ten plagues brought upon the pharaoh of Egypt.

In the Middle Ages, frogs and toads were often linked with witchcraft, leading people to connect them with evil. Frogs and toads carry a reputation as standard ingredients in potions designed for evil doings; witches were even thought to "brew the weather" by tossing snakes and toads into boiling water and invoking evil as they stirred this odd soup.

Traditionally, humans display more ill will toward toads than other anuran species. With their bumpy skin, lack of froglike grace, and tendency to be found in the dirt, toads are often viewed as ugly creatures of little value. Most children have been told touching a toad will cause warts. This folktale is completely untrue—human warts are caused by a viral infection, which anurans cannot carry—but it has contributed to the negative views humans hold toward anurans.

Such negative attitudes toward anurans have probably contributed to frog and toad decline. Humans are more likely to contribute to the conservation of animals that they like: Frogs with their slimy skin and bad reputation are not as likely to gain public sympathy and support as endangered animals considered noble, such as the bald eagle, or intelligent, such as the gorilla.

Not all human cultures dislike frogs, however. The aborigines of Australia use frogs to help them survive harsh outback conditions. As herpetologist Chris Mattison states, "Some of the desert-dwelling species, notably the water-holding frog, *Cyclorana platycephalus,* are used by aborigines as a source of water—when squeezed they yield a surprisingly large quantity of (allegedly) pure water from their bladders."[21] When unable to find water in the desert, the aborigines learn how to dig up frogs and drink the fluids.

Many cultures also depend on frogs to predict the weather. According to researcher Robert Hofrichter,

> The only European frog that truly does "announce the weather" is the tree frog. However, it does so only for weather patterns that are imminent and not for forecasts of the next day's weather. When the sun shines and it is neither too cool nor too windy, tree frogs perch a few yards/meters up on branches and vegetation; when it rains they sit just above the ground. The observation that tree frogs call out more loudly on warm, still evenings than on cold evenings is correct. [22]

Frogs are often not a reliable means of weather forecast, but were a good means of prediction before modern technology made accurate forecasting possible. By observing and listening to the tree frogs, people have been able to predict whether or not rain is on the way.

In modern times the image of frogs and toads has been popularized as an advertising tool to sell products. Frogs are

Raining Frogs and Toads: Folklore or Fact?

The stories people tell of rainstorms that include frogs and toads falling from the sky are not entirely folklore. It may seem like an impossibility, but while extremely rare, this unusual sort of storm does occur.

Reports in the Decatur *Daily Republican* in 1883, for example, described the decks of two steamers, boats moored on the Mississippi levee, observed to be covered with small green frogs. These frogs had come down with a drenching rain that had fallen the night before. Another report in 1873 in *Scientific American* magazine described a shower of frogs darkening the air and covering the ground as a result of a rainstorm in Kansas City, Missouri. Today there are still occasional reports of such occurrences. So how does it happen?

Climatologists, scientists who study weather, believe that these frog and toad rainstorms are caused by tornadoes. As the tornado passes over ponds and creeks, it picks up the small creatures with the water. As the storm moves away, the animals are lifted high into the storm. Once the winds begin to decrease, the abducted frogs drop from the sky as a rainstorm.

presented as homely but endearing or as cute character actors in the effort to sell cleaning products, beer, automobiles, and even shoe polish. Companies have adopted benign images of frogs to express to customers that their line of products is environmentally safe.

With this change in attitude, frogs have played a bigger role in popular culture: Kermit the Frog, a central character in Jim Henson's Muppets, is one of the most beloved puppet characters in generations, and environmental conservation groups have appropriated his "It Isn't Easy Being Green" message to win sympathy for various causes. Unfortunately, this heightened appreciation for and interest in frogs has put pressure on declining populations of anurans.

Jim Henson's Kermit the Frog has achieved celebrity status in popular culture and environmental organizations have used the Muppet to advance their causes.

Frogs for food

Human civilizations have included frogs in their diet throughout recorded history. For the most part, this human consumption has had little effect on frog populations. However, frogs are considered a delicacy in many parts of the world, and certain species with particularly long legs are prized. In most places people cook and eat only a frog's legs, seeking frog legs that are especially meaty. Some frog species that meet this criteria, such as the California red-legged frog, have been harvested to the point of endangerment.

Because several frog legs must be served together to make a single meal, harvesting frog legs means destroying huge numbers of frogs. According to researcher Joseph Schmuck, "In order to obtain frog legs it was customary—and still is— to sever the rear limbs from the living animals right where they were caught. Still alive, the 'refuse' is thrown back into the water."[23]

Today Europeans and Americans consume a large share of the frog legs that are consumed by humans. Originally native frog species filled the demand of these consumers. However, with the dramatic loss of amphibian habitat in Europe and the United States most edible frog species in these regions have been given protected status and may not be hunted. Today the majority of edible frogs are collected in Asia.

The quantity of frog legs harvested for human consumption in recent years is not trivial. In 1987, for example, 3,004 tons of frog legs were exported from Indonesia, representing 60 to 82 million frogs killed. A 1998 estimate put the annual consumption of frog legs worldwide at nearly 200 million pair.

Such a high demand for frog legs might seem to indicate the profitability of farming frogs for the human market. Unfortunately, few frog farmers have been successful. With so many frogs apparently available for capture in Asia, it is much more inexpensive to harvest the frogs from the wild. Farmers who raise frogs cannot produce enough frogs to pay for the process of farming them. Researcher Kathryn Phillips comments, "Perhaps after the last populations of large, wild, palatable frogs are wiped out for commerce—as they eventually are sure to be in Indonesia—prices for frog legs will become high enough that farming will be feasible."[24] Scientists agree that if the current popularity of eating frogs continues, many species of edible frog may become extinct.

Frogs for research

Humans collect frogs and toads for several uses other than as food. Most high school biology students dissect animals, meaning they cut open dead animals to examine their organs in a classroom setting in the study of animal anatomy. The majority of animals that are harvested for this purpose are wild caught animals. According to Jonathan Balcombe, an associate director of education at the Humane Society, "Approximately 10 to 12 million animals are killed for dissection exercised every year in the United States. Of these animals, I estimate that 99 percent are wild caught."[25]

Bullfrogs and leopard frogs are the main species that are used for classroom dissection. Biological supply companies are unable to breed frogs in a cost-effective manner, so these frogs are all wild caught. The frogs are captured by hand or with nets, gathering frogs in massive numbers. One species that has declined due to this practice of harvesting for dissection is the Canadian bullfrog. Millions were sold to U.S. biological houses until the population decline was noticed in 1975.

Most teachers feel that dissection is a very important part of biological studies and in many schools the dissection of wild caught animals continues to occur. This may change as more economical and humane ways to study animals appear in the classroom. Today it is possible for students to use videotapes, three-dimensional anatomical models, and interactive CD-ROMs such as "digital frog" to study anatomy and dissection on the computer. Frog enthusiasts are hoping that this trend changes the current practice of harvesting frogs for academic research. However, frogs have long played a role in laboratory research as well.

One of the most famous frogs used in a laboratory setting is the African clawed frog, *Xenopus laevis,* important in human pregnancy testing. The African clawed frog replaced the rabbit as a well-known test subject for determining if a woman was pregnant. When urine from a pregnant woman is injected into a female frog, within a few days the hormones in the urine cause the frog to lay eggs. If the woman is not pregnant, the frog does not produce eggs. Frog testing has largely been replaced by more modern test methods, but many frogs are still used in other scientific research.

The clawed frog is a very hardy species that is easy to breed in captivity, so their use in scientific research has not had much impact on their numbers in the wild. However, the qualities that make the frog excellent for laboratory use also make the frog a menace to other species of frogs. When accidentally introduced to foreign habitats the clawed frog is

A biology student dissects a frog. Recently, more humane and high-tech ways to teach students about anatomy have been introduced.

highly adaptable and impossible to control once it establishes itself in a new habitat. According to researcher Kathryn Phillips, "It can live in the dirtiest water imaginable and will eat anything that moves, including native frogs. It can do irreparable damage if let loose in a delicately balanced environment."[26] The threat posed by the clawed frog has been demonstrated by accidental releases in the United States, proving that the damage that introduced frogs can do to native anuran species is severe.

The pet trade

Frogs and toads are popular terrarium pets, but the heavy demand for the most exotic and colorful frogs has put a tremendous strain on many species of frogs. Like frogs that are imported for food, many of the species of frogs that are desirable as pets are less expensive to harvest from the wild than to breed. This means that there is much profit to be made by exporting attractive frogs from their native countries. According to researcher Joseph Schmuck, "Often small local populations are radically 'defrogged' by local gatherers who are encouraged by unscrupulous merchants (whole-sale traders). This has and continues to affect the genera *Dendrobates, Phyllobates* and *Mantella.*"[27]

Dendrobates and *Phyllobates* are poison arrow and dart frogs. These species of beautiful, brightly colored, poisonous frogs are collected from Central and South America. Although these species are not currently listed as globally endangered, they are quickly losing rainforest habitat, and biologists warn that overcollecting warrants their protection under the international agreement known as CITES.

In 1973 the Convention on International Trade in Endangered Species of Wild Fauna and Flora (CITES) was created in Washington, D.C. Currently 136 nations are convention members, incorporating CITES guidelines into their national laws. The purpose of CITES is to limit or even end trade in a species depending on its level of endangerment. The general rule is that the number of specimens removed from the wild cannot exceed the species' reproductive abilities.

Project Golden Frog: Saving a Cultural Icon

The Panamanian golden frog, *Atelopus zeteki*, is one of the most famous of the endangered frogs in the world. The frog's fame lies in its beauty and cultural significance. In Panama native peoples in the distant past believed the frog to be good luck. Present day Panamanians cherish the frog as a national symbol, an animal unique to their country. The image of the golden frog can be found on everything from T-shirts to lottery tickets. Sadly, however, its celebrity and cultural value may have played a role in its decline.

Currently the golden frog is an endangered species threatened by loss of habitat, pollution, and overcollection for the pet trade. Scientists fear that if conditions for the frog do not improve, it may become extinct within the next five years. In response a group of concerned biologists formed Project Golden Frog (PGF), a joining of scientific, educational, and zoological institutions in the Republic of Panama and the United States. PGF is concentrating on education, field studies, captive breeding, and finding financial support for the project.

Project Golden Frog currently involves fifteen zoos in their project and is having success in breeding the frogs. The captive frogs have produced several thousand offspring to date. PGF hopes that the offspring can ultimately be used to repopulate habitat throughout their former range in Panama.

Mantellas have recently been added to the list of protected species under CITES. Herpetologists have been pressuring CITES to protect the *Mantella* genus for more than ten years. Mantellas are indigenous to the Indian Ocean island of Madagascar. According to writer Wendy Williams, "No one knows how many frogs still live in the wild or how serious the collecting drain has become. The collectors themselves say that a decade ago a day's fieldwork would yield about 2,000 frogs. Today, they say, the number has dwindled down to between 100 and 150."[28] This worries scientists, especially

Mantella: A Species Disappearing Due to Collection

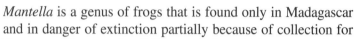 *Mantella* is a genus of frogs that is found only in Madagascar and in danger of extinction partially because of collection for the pet trade. There are at least twelve different species of this genus and they are found in a variety of habitats on the island. There are species that live on the warm, moist rainforest floor, which exists at sea level. Other species prefer hotter, drier areas or the more temperate areas in Madagascar's higher elevations. What they all have in common is their bright coloration.

Just like the poison dart frogs, which are also brightly colored, mantellas have toxins in their bodies, which make them poisonous if consumed. Scientists believe that many prey species use bright coloration to warn predators that they are toxic and eating them will have consequences. These bright colors, however, have made the mantellas attractive to the pet trade.

Scientists fear that this entire genus of frogs may disappear due to overcollecting. In 1997 more than three thousand mantellas were imported into the United States and European import numbers are even higher. Currently the frogs can still be traded across international borders, but CITES restrictions limit that trade.

The brightly colored and poisonous mantella frog is endangered partially due to collection for the pet trade.

in light of the fact that many *Mantella* species were only recently discovered. Overcollection could mean that this little known genus of frogs may disappear before scientists have a chance to thoroughly study them.

It is important to realize that human activities considered harmless for centuries—keeping frogs as pets, studying and dissecting frogs in school classrooms, enjoying a meal of frog legs—have become threats to the stability of wild populations that are under tremendous stress from other factors. These additional new problems are causing entire populations to vanish, and have caught the attention of the general public as well as the scientific community.

5

Vanishing Anurans

MANY SPECIES OF anurans are vanishing from all habitats at a rapid rate and the causes of these declines appear to be numerous. In the early 1970s field researchers started noticing rapid and apparently sudden amphibian decline. The alarm was raised in the 1980s when species such as the golden toad disappeared from habitat untouched by humans. Its disappearance could not be explained as natural extinction according to normal evolutionary processes. According to scientist Britta Grillitsch, "The natural extinction of an amphibian species should occur approximately every 1,000 years."[29]

A species is not officially declared extinct until extensive searches for the species have been conducted in all seasons and over a period spanning several generations of the animal's expected life span. Currently five anuran species are considered to have gone extinct within the past thirty years. At least nineteen other anuran species that have not been seen for many years may be extinct but as yet are classified as critically endangered until the criteria for extinction is officially met. Scientists agree that there is no one single cause of this decline; rather, anurans are suffering from a combination of problems that are only recently identified and for which solutions have yet to be found.

The chytrid fungus

The chytrid fungus (pronounced ki-trid), *Batrachochytrium dendrobatidis,* is a long-identified fungus, commonly found in Earth's environment, that was until recently known to affect only plants and insects. Suddenly, however, this fungus has turned lethal to amphibians. Frogs are the first vertebrates known to have been harmfully affected.

The chytrid fungus feeds on keratin, the substance composing human nails and hair. Tadpoles have keratin in their mouthparts. The chytrid fungus infects tadpole keratin but because tadpole keratin is localized, the tadpoles survive to adulthood. Only when they metamorphose into adult frogs, developing much more keratin in the process, is the fungus capable of sickening and killing the frog.

Scientists are uncertain precisely how the chytrid fungus kills adult frogs. Some believe that the fungus may secrete a toxin that ultimately kills the frog. Other researchers believe that this fungus invades the top layers of the frog's skin, causing damage to the keratin layer on the skin surface. Since frogs drink and breathe through their skin, the fungus may kill the frogs by disrupting these mechanisms.

Tadpoles swim about a pond. The chytrid fungus attacks the mouthparts of tadpoles, but it is not deadly until the tadpoles reach maturity as frogs.

Another uncertainty is where the fungus came from. Currently it has appeared on six continents and is spreading rapidly in Central, South, and North America as well as Africa, Europe, New Zealand, and Australia. Some herpetologists speculate that the fungus survives on the soles of shoes and is spread to wildlife habitat on the boots of tourists. In Australia it is believed that the fungus arrived on the skin of a foreign frog that was delivered in a fruit box. Some scientists argue that the chytrid fungus has been present in anurans all along, suggesting that something is lowering normal resistance to the fungus in the amphibian immune system. Scientist Don Nichols points out, "It doesn't do a parasite any good to kill its host; other factors must be tipping the balance."[30] What scientists are sure of is that the fungus is lethal and has proven to be especially destructive in Australia, where the destruction of a research colony first sounded the alarm.

 Gastric Brooding Frogs: A Victim of the Chytrid Fungus?

The first gastric brooding frog, *Rheobatrachus silus,* was discovered in southeast Queensland, Australia, in 1973. The newly discovered frog astounded scientists in the lab when a female was found giving birth to live young from its mouth. It was found that females brood their young in their stomach without digesting them. The new species of frog created quite a stir in the scientific community.

The second gastric brooding frog, *Rheobatrachus vitellinus,* was discovered in 1984 on the central coast of Queensland. Both species have now disappeared in the wild and scientists believe that *R. silus,* at least, is now extinct (there are none in captivity). Scientists are unsure why the two species have disappeared from the wild. Herpetologists are still studying captive *R. vitellinus,* but it has not been seen in the wild for several years. Some researchers believe that this species of frog holds the secret to curing stomach ailments, such as ulcers. Scientists believe that the frog may hold clues to controlling the production of stomach acid. The permanent loss of the gastric brooding frog may be a loss for not only herpetologists but medical researchers as well.

Herpetologist Michael J. Tyler's mid-1970s discovery of the gastric brooding frog in Australia made headlines around the world. Scientists were intrigued by the species' ability to allow its young to develop in its stomach. According to writer Ron Cowen, "Tyler's report of this unusual species made headlines in the mid-1970's, when the creature was so abundant in the rain forest near Brisbane that researchers could observe 100 of them in a single night. But by 1981, the brooders had vanished."[31] At the time Tyler was unsure what was happening, but the gastric brooders were not the only frogs to suddenly suffer declines. As the trend continued, it was apparent that something was wrong with the frog populations.

Most of the declining species come from pristine, or untouched, habitats at higher elevations, which scientists assume remain unaffected by humans. Originally this was a puzzle, but eventually chytrid was named as the culprit. The fungus was first identified in 1998, but research now shows that the fungus had existed in Australia since at least 1978. There are now forty-six species of anuran affected by the chytrid fungus in Australia. Herpetologists are working hard to research this infection, fearing that chytrid may cause the extinction of many Australian species of frog which, like the gastric brooding frog, can be found nowhere else in the world. Almost half of the affected species come from this continent. This trend may change as chytrid spreads through other continents.

In 1997 researcher Karen Lips found dead and dying frogs littering her research site in Panama. Lips commented, "I've been going to Fortuna, Panama since 1993. In 1997, I returned to find all these dead frogs. They looked fine, like they went to sleep and didn't wake up."[32] Lips collected the bodies so that they could be examined for cause of death. The cause was chytrid. Just like in Australia, the fungus had affected frogs in the tropical forests in high elevations of Central America.

Researchers soon found chytrid in the United States. Leopard frogs from Arizona exhibited the infection and chytrid has now been implicated in the decline of the boreal toads of the Colorado Rocky Mountains in the 1970s. Cynthia Carey, who was studying the boreal toads for a doctoral dissertation, watched as all the toads in her study mysteriously died: "In the

1970's we lost 85 to 95 percent of all Colorado frog and toad populations at high altitudes. If you estimate how long it would take for populations to come back from that kind of mass mortality, the answer is: several hundred years."[33] Today more North American frog species are being discovered with chytrid infections and scientists are unsure how to stop the fungus or treat infected frogs. Carey points out that "now that we know the chytrid fungus is the cause, it isn't any less frightening. We know what it is, but we can't stop it."[34]

Chytrid research

Around the world herpetologists and other scientists are researching various aspects of amphibian chytrid. Tests in the laboratory have shown that 99 percent of the frogs exposed to chytrid die. Researchers in the field are taking care not to spread the fungus from one frog population to another, while trying to find more information about the frogs and this infection. The Australian Animal Health Laboratory is currently culturing, or growing, the fungus and testing antifungal drugs in an attempt to find a cure. Even in habitats where the frogs have disappeared due to chytrid, the fungus remains: Until scientists can discover how to make the frogs immune or remove the fungus from the area, the frogs cannot recover and captive populations cannot be released back into the wild.

Other scientists are concentrating on studying frogs that have remained unaffected by chytrid infections in their habitat, trying to understand the preferences of the fungus in hopes of discovering a way to control it. Researcher Karen Lips states, "At the rate chytrid is going we may lose all the frog species in every area above 500 meters [1,600 feet]."[35] Scientists believe that without research, this fungus cannot be understood and many more species of anuran may become extinct.

Scientists are unsure why this naturally occurring fungus has become lethal to frogs. Some herpetologists speculate that human-induced factors, particularly environmental pollution, are causing frogs' immune systems to fail and leaving frogs vulnerable to diseases to which they were previously resistant. This theory has been applied to other recently discovered frog and toad problems.

Amphibian mutations

In 1995 Minnesota school children on a field trip to a local pond found deformities in almost half of the leopard frogs that they caught. Most of the deformed frogs had extra limbs or missing limbs. Once the children's discovery was announced through the media, reports came flooding in from other regions of the country as people began to notice other local frog deformities. This raised an alarm with herpetologists around the country who feared that the deformities could contribute further to amphibian decline. Clearly, frogs with deformed limbs were unable to effectively hunt or escape predators.

Currently, fifty-two species of amphibians have been reported with these deformities in forty-six states in the United States as well as four Canadian provinces. Baffled scientists focused their research into causes on three major suspects: pollutants, ultraviolet light, and parasites, all of which are currently considered factors in other forms of amphibian decline.

This frog's four-legged body is the result of a mutation. Many scientists believe these are the result of microscopic flatworms inside the bodies of tadpoles.

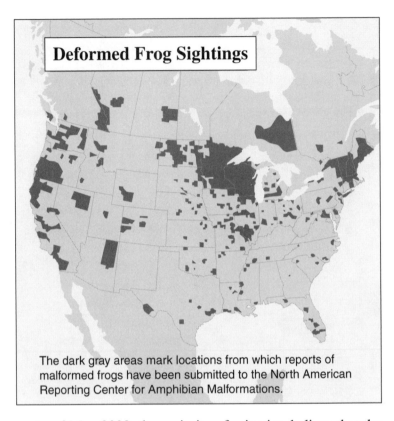

Deformed Frog Sightings

The dark gray areas mark locations from which reports of malformed frogs have been submitted to the North American Reporting Center for Amphibian Malformations.

As of May 2002, the majority of scientists believe that the frog deformities are caused by a microscopic flatworm also known as a trematode. The trematodes, *Ribeiroia ondatrae,* are commonly found in ponds and are carried by snails. Tadpoles that share ponds with infected snails pick up trematode larvae, called cercariae. The larvae enter the frog through its skin and congregate in small balls, causing cysts, which interfere with the tadpole's metamorphosis into a frog. When the cysts form in tissues that develop into new limbs, they can cause duplicate or missing legs.

The trematode can also affect humans, but with much lesser consequences. A rash called "swimmer's itch" is caused by the same parasite in humans, contracted while swimming in ponds and lakes. However, the human immune system is able to defeat the parasite, leaving the infected person with nothing more than itchy skin.

Many scientists point out that this parasite has always existed in these frog habitats, and the deformities are not new.

Indeed, frog deformities have been reported since preindustrial times. Professor of biology and herpetologist Mike Lannoo states, "The thing we have to be careful about with amphibians is making blanket statements. We lack the long term research data to make the type of conclusions that we want to."[36] Most scientists agree, however, that this rash of deformities is abnormal. They also believe that something else must be occurring to intensify the effects of the trematodes in frog populations. Many researchers suggest that pesticides may be a contributing factor.

Pollution

Researchers studying the extent of amphibian deformities discovered that when they counted the number of cysts on tadpoles, they found higher levels of cysts in the tadpoles that had been exposed to pesticides. Scientists also took blood samples and found fewer white blood cells, the cells concerned with fighting infection, in the exposed tadpoles. According to researcher Joseph Kieseker, "The tadpoles that were exposed to pesticides had fewer of this particular kind of white blood cell compared to the tadpoles that we did not expose to pesticides, suggesting that pesticides make these animals more susceptible to parasitic infections."[37] Pesticides may increase the number of deformities in frogs, but scientists also believe they can have other adverse effects.

Amphibians, with their permeable skin, are perhaps the animals most easily affected by pollution in the water. At least 250 chemical compounds are known to have toxic effects on wildlife species, including anurans. Research into the specific effects of these dangerous compounds is incomplete, but scientists know that many compounds find their way into the water supply and agree that water pollution is a large factor in amphibian decline.

Pollutants arrive in anuran water sources in a variety of ways. The main source of chemical pollution is from farmland runoff, containing pesticides used to protect crops. Pesticides that are sprayed over crops may not only leach into the water system, but can also be carried in the air from farmland to ponds. These forms of pollution may not directly kill adult frogs, but can lower the breeding success of the anurans inhabiting the water, and ultimately have a negative effect on the population.

Research study findings released in April 2002 show that the common weed killer Atrazine interferes with the sexual development of frogs. Atrazine-exposed male frogs have reduced levels of testosterone, which is a hormone required in male development. These males have ovaries, female reproductive organs, and much smaller vocal organs than normal. Studies show that Atrazine could cause deformities in frogs at levels as minimal as three parts per billion, a level far be-

A California farmer sprays snow pea crops with pesticides. Pesticides have been proven to cause abnormalities and deformations in amphibian populations.

low what is considered safe for human consumption. Currently, the National Resources Defense Council is trying to have the pesticide banned.

Research has also shown that fertilizer levels that are considered safe for human consumption can kill some species of frogs and toads. At Oregon State University researchers discovered that when tadpoles were raised in water with low levels of nitrates, a major ingredient in fertilizers, the tadpoles ate less, developed deformities, and suffered paralysis, eventually dying.

The research in Oregon proved that fertilizers are a factor in the decline of the protected Oregon spotted frog, *Rana pretiosa*. The Oregon spotted frog has largely disappeared from its historical range in the heavily farmed Willamette River Valley. In these heavily agricultural areas, nitrates from the fertilizers run off into the local ponds and, even at levels considered safe for human drinking, poison tadpoles. Fertilizer runoff into ponds has also been implicated in trematode infestations.

The nitrate in fertilizers feeds the aquatic algae that is the basis of the food chain in ponds. As the algae increases, so does the number of snails that feed on the algae and carry the trematodes. The more snails there are, the more trematodes there are, and the higher rate of infection for frogs. Pesticides and fertilizers thus not only disrupt the ecological balance of water bodies but encourage the growth of the agents that are killing frogs and toads.

Other unnatural changes to the water can affect anurans as well. Acid rain, for example, has been implicated in the decline of some European species of frogs. Acid rain is rain that is saturated by sulfur and nitrogen, chemicals that build up in the atmosphere due to the burning of fossil fuels such as coal and petroleum. This rain raises the acidity of small ponds as it falls into them. Higher acidity can have adverse effects on amphibian reproduction.

Acidic water can not only cause deformities during development, but can also decrease the chances for anuran survival. Eggs that are in acidic ponds can lose their resistance to infection. The acid destroys the gel that surrounds the eggs, allowing bacteria to attack the eggs. Tadpoles that have been compromised by acidic water are also more likely to be eaten by predators.

Scientists are unsure if acid rain has affected frog populations in North America, but acid rain has been proven as a factor in the extinction of local amphibian populations in Scandinavia in the 1970s. Herpetologists believe that although it has not been proven as a cause of decline in other countries, overall, air pollution is an important factor to study in anuran decline.

Some scientists theorize that due to loss of ozone, the layer in the atmosphere that blocks ultraviolet B (UV-B) radiation, anurans are being exposed to unhealthy levels of UV-B and this is contributing to frog declines. The loss of ozone is an effect of the release of industrial chemicals such as chlorofluorocarbons (CFCs). This lessening of the ozone allows in ultraviolet radiation, which is a high energy radiation and has the capability to damage living organisms. Recent studies show that exposure to UV-B can cause lowered immunity in frogs and in their eggs.

Scientists exposed anuran eggs to UV-B first in the laboratory in order to discover how the radiation might affect them. As they expected, their research showed that amphibian eggs and embryos did not develop properly when exposed to UV-B. The next step was to design and conduct field studies to determine the effect of UV-B on wild populations.

Herpetologists are most concerned about UV-B in high altitudes where the atmosphere is thinner and there is less protection from UV-B. Significantly, the chytrid fungus is most severely affecting frog species at these high altitudes. Scientists speculate that UV-B somehow stresses these frogs' immune systems, lowering their resistance to infection. Researcher Kathryn Phillips states that in one study conducted by scientist Andrew Blaustein at a mountain lake, "One-fourth of the Cascade frog and western toad eggs died when they were not shielded from UV-B radiation."[38]

A recent study in 2002 sponsored by the U.S. Environmental Protection Agency showed that UV-B was a harmful factor in some anuran habitats. Twenty-six wetland areas

were measured for UV-B radiation and three sites were shown to pose a risk to amphibians living in the area. This could also mean that UV-B lowers immunity at lower elevations as well, allowing frogs to be more easily infected by trematodes, resulting in anuran deformities. Although UV-B may not be the main factor contributing to anuran decline, many scientists feel it is one factor in a chain of problems that is causing the decline of anurans around the world. Faced with significant species decline, the scientific community has turned its efforts to conservation.

A red-legged frog is poised to hunt. Scientists have promoted conservation efforts as a result of the population decline of anurans around the world.

6

The Future of Frogs and Toads

SCIENTISTS ALARMED BY the dramatic decline of anuran species are searching for ways to preserve the order, believing that the loss of frogs and toads can negatively affect all other species. Researchers and frog supporters around the world are currently seeking ways to help frogs survive in the wild and ways to gain public support for frog conservation.

Why conserve frogs?

Many scientists feel that frogs are "biomonitors" of the environments they live in and recognition of this important role should lead people to get involved in their conservation. A biomonitor is a plant or animal that is easily studied and that quickly shows signs of distress when a habitat is compromised. Frogs and toads meet these qualifications; they are easy to catch and very sensitive to subtle changes in their specific habitat.

Researchers believe that the decline of frog species may indicate problems in the frogs' habitats that could eventually harm humans as well. Frogs drink and breathe through their skin, which means that toxic substances can move freely into their bodies. Toxins in the environment, such as the various forms of water and air pollution, quickly concentrate in frogs' fatty tissues and are easily detected by scientists as well as manifested in symptoms of disease in the animals.

Environmental scientists believe that by watching frog populations they are better able to take action and correct environmental problems early on. Other species may be sensitive to environmental changes, but are not as easy to study as anurans. Herpetologist Chuck Peterson states,

> I work with reptiles too, but they're much harder to study than amphibians. They don't all get together in a pool to breed. They don't call. You can't look into a pool and see their larvae. Amphibians are probably more susceptible to changes in the environment, but equally important is the fact that you can see what's happening to them. Amphibians are a good indicator of ecosystem health.[39]

However, many other scientists argue that frog declines are not an early warning, but a warning that comes too late. Once the frogs start to decline, the damage has already been done.

Herpetologists point out many other ways anurans are beneficial. Adult frogs and toads are excellent predators and voracious eaters. Since anurans feed mainly on insects, they

Frogs and toads are especially valuable creatures because they keep insect populations under control. Here, a Bufo *devours a dragonfly.*

A snake swallows a frog whole. A decline in the number of frogs and toads can adversely affect the survival of their predators.

can keep insect populations under control. A single toad can eat as many as ten thousand insects a year. In the absence of frogs undesirable insect populations could greatly expand: In Asia, for example, where the majority of frogs are harvested for the food industry, the decline in frog numbers has been linked to the increase in mosquitoes and other insects.

The lack of frogs could also affect the animals that depend on frogs as a food source. When frogs disappear from a habitat, predators that feed extensively on anurans will be affected. The animals that previously fed on frogs will have to rely on a different food source. If the alternate food source is not sufficient, the predator species will decline as well.

In the Sierra Nevada mountain range of the United States, frog declines have been linked to the decline of the mountain garter snake, *Thamnophis elegans elegans.* The snake feeds primarily on the mountain yellow-legged frog, *Rana muscosa,* and the Yosemite toad, *Bufo canorus,* both of which are disappearing from their former habitats. If the garter snake population diminishes, it may also cause declines in birds of prey and other predators that primarily feed on the snakes. Research continues into the effects of missing anurans from the environment, but many scientists feel that frog declines can upset the balance of predators as well.

Herpetologists also point out that the disappearance of frogs and tadpoles can upset the quality of the habitat. Anurans shape the habitats that they are a part of and tadpoles are an important part of the overall health of a pond system. Many species of tadpoles feed primarily on the algae that grow naturally in freshwater ponds. They also feed on the decomposing plant matter at the bottom of the pond. In this way, tadpoles control the quality of the pond they live in, making it suitable for other species that live there.

When tadpoles are no longer a part of the habitat, algae growth is unchecked. In Fortuna, Panama, where the majority of the frogs have disappeared due to chytrid, many streams are now choked with algae. Scientists are unsure exactly how this will affect fish, birds, and other animals that share the habitat.

Potential medical benefits

In recent years scientists have discovered that the poisons that some anurans exude for protection are chemicals that could prove to be medical treasures. Recent research has shown that the molecules called peptides secreted by some frogs have the potential to reduce high blood pressure, stop blood clotting, destroy bacteria that are resistant to treatment by conventional antibiotics, limit bone marrow damage during chemotherapy, and make crops resistant to insect attack. Professor Chris Shaw at the University of Ulster in Ireland states, "Biological warfare has been going on in the rain forest for millions of years as each organism living there has fought for its survival. I believe that we can put that kind of biological weaponry to use for the good of mankind in the

Wood Frogs: The Future of Freezing Human Cells?

There are several species of frog that are able to freeze in the winter and not die, thawing in the spring to breed. One such species is the North American wood frog. This particular species is even found in Alaska where winters can be particularly harsh.

As the temperatures lower with the approach of winter, the frogs burrow under leaf litter or in the mud at the bottom of ponds, where they are somewhat insulated from the freezing air above. The frogs still become very cold, but are able to survive as long as they do not get colder than twenty degrees Fahrenheit. That is more than ten degrees below freezing.

The wood frog goes into a deep hibernation, its breathing and heartbeat slowing and eventually stopping. As much as 65 percent of the water in its body slowly turns to ice. In most animals, this would destroy cells and kill the animal. In the wood frogs, glucose, a type of sugar, is manufactured in the frog's body and stops vital cells from completely dehydrating as they freeze. So although the frog's eyeballs and brain are so frozen that they become rock hard, the frog is not dead. When the environment warms, the frog revives and begins searching for the breeding ponds.

Scientists are currently studying the wood frogs to learn how they are able to freeze and recover. It is possible that if the frogs' secrets are uncovered, scientists may be able to use the same methods to freeze human cells, thawing them for future use.

Scientists study wood frogs in hopes of discovering ways to freeze human cells without damaging them.

ultimate defeat of those diseases which have thus remained intractable."[40] As Professor Shaw researches the potential of anuran toxins, human trials are currently underway in Europe for an exciting new drug derived from a poison arrow frog.

A poison arrow frog from Ecuador, *Epipedrobates tricolor,* is the source of a new painkiller. Researchers from Abbott Laboratories in Chicago have developed a new painkiller called ABT-594, a chemical derivative of epibatidine, which is a compound found on the frog's skin. ABT-594 blocks pain two hundred times more effectively than morphine. Unlike morphine, ABT-594 does not hinder respiration or slow digestive movement, and it is not addictive. Scientists in many disciplines now point out the possible benefits of studying frogs and discovering new species, hoping that public and scientific interest may spur frog conservation.

Conservation efforts

Herpetologists agree that just as there are many causes of anuran decline, there must be many different efforts launched in order to preserve endangered frogs. Most of these efforts must begin at a local level.

In order to save local species, people must become interested in anurans and inspired to take action. Local efforts can be simple. In many places, for example, programs are in place to help frogs and toads make the trip to nearby ponds to breed. Many frogs and toads are killed by automobile traffic as they cross roads by the hundreds to reach their breeding grounds. In Davis, California, people built toad tunnels beneath roadways to insure the toads have a safe trip. The Davis project did not appear to be successful—local frogs and toads did not use the toad tunnels in greater numbers, but frog advocates in other places were inspired to look for other ways to help frogs make a safe crossing.

Toad tunnels like this one are constructed by conservationists to allow anurans to safely cross busy roads.

In England many toad lovers are working together to help the rare natterjack toads cross the road. At night at around 11 P.M., when many English pubs close, traffic can be heavy on roads where the natterjacks are crossing to reach their breeding ponds. Millions of toads are killed in this manner. Just before pub closing, groups of volunteers go out with buckets, gathering the toads and giving them a safe ride to the pond. Other groups take conservation a step further and create habitat for diminishing species.

Gardening clubs suggest their members attract toads to their gardens to help control garden pests. Toads eat almost all of the different kinds of insects that destroy gardens. Gardeners build small ponds near their vegetable gardens to attract toads. They also give the toads a place to stay safe and cool during the day, using mulch for them to burrow under, terra cotta "toad houses," or bushy shade plants. The toads help control the insects and the gardeners help supply toad habitat.

Some groups are creating even larger areas of habitat. Giving anurans restored or new habitat can make a positive difference to frog populations. Many groups are restoring wetlands by removing invasive plants, building dams and drains, and increasing water flow. Frog advocates hope that these pockets of wetlands will help preserve local populations of frogs.

Education programs that parallel ongoing habitat and restoration projects can help efforts to save endangered species succeed. This has proven to be true with the Puerto Rican crested toad Species Survival Project (SSP). Puerto Rican students working on a school project discovered the toad, which was previously thought to be extinct. Their interest and efforts increased local awareness, which expanded conservation efforts and showed scientists that it is important to include public education in SSPs. Bob Johnson, curator of reptiles and amphibians at the Toronto Zoo, points out, "Ultimately, it is those who live in threatened habitats and with threatened species who will make the choice regarding the quality of life they wish for their successors and the wildlife species that share their landscape."[41] Public education may raise interest and encourage people to take part in conservation.

Frog counts

Scientists, believing that research is crucial to saving local species, have begun volunteer frog counts to help them gather research data. One such project is Frogwatch USA, which is a long-term frog and toad monitoring program coordinated by the U.S. Geological Survey Biological Resources Division. Frogwatch USA recruits volunteers to complement other ongoing local, national, and global amphibian monitoring efforts.

Frogwatch USA educates volunteers in locating frogs and toads and identifying their distinct calls. Volunteers learn the life cycles and habits of local species, then, directed by Frogwatch staff, go out on their own to a site registered with Frogwatch USA to monitor frog populations. Volunteers report

 ## The Green and Golden Bell Frog: Olympic Mascot

The threatened Australian green and golden bell frog, *Litoria aurea,* gained global attention when it was found on the future site of the Sydney 2000 Olympic Games. The population is one of only twelve known breeding colonies of this species of frog. Sydney had committed to environmental guidelines and when organizers discovered a population of green and golden bell frogs in an abandoned brick pit on site, they decided to preserve the frogs. With this conservation effort, the frog made headlines and became a mascot. Saving the endangered frog enhanced Sydney's image and the beautiful green and gold frog made an effective advertising image.

In the 1960s the green and golden bell frog was an abundant species. Declines of the species were noticed in the 1970s and became severe in the 1980s. Once widespread throughout the east coast of Australia, the frogs are now limited to isolated pockets. In 1995 the green and golden bell frog was listed as endangered in Australia.

Today the population at the Olympic site is flourishing. It is currently one of the largest populations in the country. There have also been sightings of frogs elsewhere in the area, indicating that the frogs' numbers continue to grow.

their results to Frogwatch USA and the information is entered into a database that is available on the Internet. Scientists at the U.S. Geological Survey believe that citizen scientists will not only help gather important information, but also raise awareness of the difficulties facing anurans and other amphibian species.

Captive breeding programs

Volunteers can help scientists research highly visible frogs, but scientists also need help studying frogs that are rare and need sophisticated study. Frog species that are difficult to

 Puerto Rican Crested Toad: An Anuran Assisted by the Species Survival Program

The Puerto Rican crested toad, *Peltophryne lemur,* is a recently rediscovered species with an uncertain future. The toad was rediscovered in 1967 by a group of high school students who were members of a teacher-led conservation group. While attending a lecture, the students heard of the extinction of the Puerto Rican crested toad and thought that they had seen the species at a local farm pond. The existence of the "extinct" species was confirmed and scientists began to search for other sites where the toad might be found.

There were only two sites found where the toads were living. A northern population near Quebradillas was identified, however no toads have been seen there since 1992. Currently the toads are found breeding only in a single pond in Guanica Forest Reserve on the southern coast of the island. The population is thought to consist of about three hundred individuals. This small number of isolated toads makes the Puerto Rican crested toad one of the most endangered of all anurans.

In an effort to save the species, the American Zoo and Aquariam Association (AZA) began the first Species Survival Program (SSP) for an amphibian. The Puerto Rican crested toad SSP was created in 1984 and focuses mainly on captive breeding and reintroduction. A small group of toads was taken from the wild and distributed into breeding projects in various zoos.

Coordinated by the Toronto Zoo, there are currently twenty-two zoos that have toads, including several with breeding populations. As of December 2001 there were 294 individuals in captivity. Researchers continue to study the toad both at zoos and in the wild, hoping to discover ways to ensure the species' survival.

study in the wild are sometimes brought into captivity where captive populations can be studied and possibly reintroduced into the wild.

When populations fall to levels so low that scientists believe the species is near extinction, specimens are often brought into captivity to be placed into breeding projects in zoological institutions. Scientists hope that the problems facing the frogs can be solved and that the species can then be reintroduced to its previous habitat.

Reintroduction from captive breeding has several problems that must be overcome. If the habitat in which the species previously thrived is not available or has been corrupted, reintroduction is not going to be successful. The frogs must have untainted habitat in which to live. Frogs brought into captivity and rereleased must also be free of disease. In captivity frogs are quarantined, or separated from other frogs and evaluated for diseases, but upon reintroduction may still spread disease to other wild species. Sam Lee, senior keeper at the Wildlife Conservation Society (WCS) and coordinator for the Kihansi spray toad project, states, "Once you bring an animal out of captivity, there is no guarantee of quarantine."[42] Even if reintroduction is not a possibility, herpetologists agree there is much to learn from studying captive frogs.

Studying anurans in captive populations bred in zoos may give herpetologists a better understanding of the species' needs in the wild. Many species are secretive and difficult to study in the wild. Within an aquarium in a zoo, scientists can watch and learn the natural history of a species.

Studying the mountain chicken in captivity gave scientists a better understanding of how best to preserve the species in the wild. Researcher Kevin Buley states, "Work at the Durrell Conservation Center at Jersey Zoo [British Channel Islands] with a captive breeding programme for the species, has led to a unique understanding of how the species breeds."[43] Watching the mountain chicken revealed that the females take unusual care of their offspring, repeatedly laying unfertilized eggs to feed tadpoles while in their burrow. This meant that if the mountain chicken females were harvested for human consumption during breeding season, the tadpoles would not

A Central Park zookeeper poses proudly with a giant toad. Conservation efforts by zoos help to ensure the survival of anuran species.

survive. Such information allows scientists to make recommendations for conservation programs. This need for research has led zoos around the world to expand their herpetology departments.

In 2000 the Detroit Zoo opened the National Amphibian Conservation Center (NACC), a $6.1 million, twelve-thousand-square-foot facility where herpetologists study amphibians and visitors learn about amphibian decline. NACC participates in nearly a dozen different endangered frog–breeding projects and has had success in breeding many of the species.

Scientists believe that zoos are important not only for research, but can also be an "ark" for species that are disappearing in the wild. If a species goes extinct in the wild, humans may still be able to keep them alive in captivity and discover a way to save them. However Kevin Zippel, curator at NACC, warns that "conservation has to happen in the field. Captive zoo populations are merely temporary lifeboats."[44] Yet field researchers such as Karen Lips who are facing the immense declines caused by chytrid feel that zoos may be the only hope for some species. Lips states, "We are running

Wyoming Toad: A Captive Breeding Success

The Wyoming toad, *Bufo baxteri,* was discovered in 1946. At that time the toad was abundant in the wetlands and irrigated meadows of Wyoming's southeastern plains. However, by the 1970s the species went through a drastic decline and was confined to privately owned lands surrounding Mortenson Lake. U.S. Fish and Wildlife Service listed the Wyoming toad as endangered in 1984. In an effort to protect the last population of toads, the Nature Conservancy purchased the lake and surrounding lands of about eighteen hundred acres. Despite this effort, the species continued to decline and by 1994 the toad was extinct in the wild.

Fortunately, there was a population of the toads in captivity. In 1996 the American Zoo and Aquarium Association (AZA) approved a Species Survival Program (SSP) for the toad. Working together with state and federal authorities, AZA began a breed-and-release program. Populations of the Wyoming toad were housed at eight different zoological institutions, where they were bred for future release. By 1996 approximately ten thousand toads had been reintroduced in previous habitat. Very few of the toads seem to have survived to breeding age, but researchers heard breeding calls in 1998 and are hopeful that the species will be able to reestablish itself with the continuation of the SSP program.

Considered extinct in the wild, ten thousand Wyoming toads were reintroduced in 1996 due to the enactment of a Species Survival Program.

out of time and we don't know how to help the frogs in the wild. Bringing them into captivity might help us solve the problems by allowing us to better research them."[45]

Although the pet trade threatens mantellas and poison arrow and dart frogs, many zoological institutions and serious amateur herpetologists have successfully bred these species in captivity. As these captive-breeding programs become more successful, captive-bred anurans will become less expensive and pet traders and pet owners will be more likely to buy frogs that have been bred in captivity over wild specimens. If there is no demand for harvesting frogs from the wild, that trade in frogs will cease. This may be important to the survival of anurans in the wild considering that amphib-

The American Zoo and Aquarium Association Species Survival Program

The Species Survival Program (SSP) began in 1981 as a joint effort between members of the American Zoo and Aquarium Association to manage populations of selected endangered species in zoos and aquariums in North America. Each SSP manages the breeding of a species in order to maintain a healthy and self-sustaining population of animals.

The SSP plays an important role in the conservation of endangered species. It designs the "family tree" of a particular captive population in order to achieve maximum genetic diversity, meaning that the breeding pairs are not closely related to each other. Breeding and other management recommendations are made for each animal and transfers between institutions are carefully planned. The SSP for each species considers all of the factors involved in a species' care, sometimes even recommending against the breeding of specific animals so as to avoid having the population outgrow the holding space.

SSPs also develop husbandry manuals, which describe the best care and diet for the species in captivity. These standards make it easier to evaluate problems in a breeding population and also make transferring from one institution to another a simpler transition.

An important part of the successful operation of an SSP is the studbook. A studbook contains vital records of an entire captive population of a species, including deaths, transfers, and lineage. With appropriate computer analysis, a studbook enables zoo managers to make sound management and breeding decisions regarding a population. Currently, the Wyoming toad and Puerto Rican crested toad are part of the SSP program.

ians around the world are diminishing and no one can pinpoint a single cause for their decline.

The future is uncertain

Scientists are unsure what the future of anuran species will be. They agree that amphibians overall are in a state of decline. However, some scientists argue that this decline is no more severe than that of other species. After all, in 1995 there were 346 species of fish in North America considered endangered or threatened and about 1,000 species of birds worldwide considered at the same status. Still, herpetologists point out that the anuran decline only seriously began in 1989 and has been rapidly progressing since. This dramatic progression is sufficient to cause alarm.

Herpetologists feel that research is essential to saving anurans. The causes of decline seem to be numerous and must be better understood before effective countermeasures can be proposed. Dealing with the little understood chytrid fungus is a case in point. Studies are needed not only by herpetologists but in immunology and other disciplines in order to solve the problems facing frogs today. Researcher Karen Lips states, "We need more people to understand what's going on. We need all sorts of scientists besides herpetologists to look at these problems, scientists like virologists and immunologists."[46] Research may uncover solutions to the many problems facing frogs and toads, but scientists all agree that the single most important thing that can be done to preserve failing anuran populations is to protect their habitat. Proper habitat, free of disease and pollution, may ultimately be what saves many species of frogs and toads.

Notes

Chapter 1: Frogs and Toads: The Order of Anurans

1. Chris Mattison, *Frogs and Toads of the World.* Dorset, UK: Blandford Press, 1987, p. 11.

2. Quoted in Roy W. McDiarmid et al., *Tadpoles, the Biology of Anuran Larvae.* Chicago: University of Chicago Press, 1999, p. 215.

3. Quoted in McDiarmid, *Tadpoles,* p. 24.

Chapter 2: Declining Habitat

4. Michael J. Tyler, *Australian Frogs and Toads, a Natural History.* Ithaca, NY: Cornell University Press, 1994, p. 88.

5. Susan Tweit, "Spadefoots and Termites," *Southern New Mexico Online.* www.southernnewmexico.com.

6. Quoted in Tyler, *Australian Frogs and Toads,* p. 91.

7. Trevor Beebee, *Ecology and Conservation of Amphibians.* London: Chapman and Hall, 1996, p. 133.

8. Quoted in Robert Hofrichter, *Amphibians.* Buffalo: Firefly Books, 2000, p. 129.

9. Wendy Williams, "Flash and Thunder (Frogs of Madagascar)," *Animals,* July 2000. www.findarticles.com.

10. Jason B. Searle, "Hey, Who Turned the Water Off?" *Wildlife Conservation,* vol. 104, no. 1, 2001, p. 11.

Chapter 3: Threats to Survival in the Wild

11. William E. Duellman and Linda Trueb, *Biology of Amphibians.* Baltimore: Johns Hopkins University Press, 1994, p. 244.

12. Duellman and Trueb, *Biology of Amphibians,* p. 268.

13. Quoted in Hofrichter, *Amphibians,* p. 176.

14. Duellman and Trueb, *Biology of Amphibians,* p. 245.

15. Duellman and Trueb, *Biology of Amphibians,* p. 258.

16. Quoted in Beth Livermore, "Amphibian Alarm: Where Have All the Frogs Gone?" *Smithsonian,* vol. 23, no. 7, 1992, p. 7.

17. Ashley Mattoon, "Amphibia Fading," *World Watch,* vol. 13, no. 3, 2000, p. 12.

18. Quoted in Rachel Nowak, "Leave Well Alone," *New Scientist,* September 2, 2000, p. 14.

19. Marty Crump, *In Search of the Golden Frog.* Chicago: University of Chicago Press, 2000, p. 283.

20. Kathryn Phillips, *Tracking the Vanishing Frogs.* New York: St. Martin's Press, 1994, p. 158.

Chapter 4: Threats to Survival Posed by Humans

21. Mattison, *Frogs and Toads,* p. 143.

22. Hofrichter, *Amphibians,* p. 207.

23. Quoted in Hofrichter, *Amphibians,* p. 215.

24. Phillips, *Tracking the Vanishing Frogs*, p. 90.

25. Quoted in Jack Rosenberger, "Harvest of Shame: Dissection's Deadly Toll Hits Frogs the Hardest," *E: The Environmental Magazine,* July/August 1998, www.findarticles.com.

26. Phillips, *Tracking the Vanishing Frogs,* p. 105.

27. Quoted in Hofrichter, *Amphibians,* p. 239.

28. Williams, "Flash and Thunder."

Chapter 5: Vanishing Anurans

29. Quoted in Hofrichter, *Amphibians,* p. 228.

30. Quoted in Mattoon, "Amphibia Fading," p. 12.

31. Ron Cowen, "Brooding over Australian Frogs," *Science News,* March 3, 1990, p. 142.

32. Quoted in Kim Y. Masibay, "Rainforest Frogs: Vanishing Act? Frog Populations Around the World Are Dying Off Mysteriously. Can Scientists Save Them?" *Science World,* March 11, 2002, www.findarticles.com.

33. Quoted in Michon Scott, "Where Frogs Live," *Goddard Space Flight Center,* March 5, 2001. http://earthobservatory. nasa.gov.

34. Quoted in Scott, "Where Frogs Live."

35. Karen Lips, professor and researcher at the Southern University of Illinois at Carbondale, phone conversation with author, September 18, 2002.

36. Quoted in Alexandra Ravinet, "Listening Hard for Frogs," *ABCNews.com,* July 2, 1999, http://more.abcnews.go.com.

37. Quoted in Stentor Danielson, *National Geographic News,* July 9, 2002. http://news.nationalgeographic.com.

38. Phillips, *Tracking the Vanishing Frogs,* p. 135.

Chapter 6: The Future of Frogs and Toads

39. Quoted in Scott, "Where Frogs Live."

40. Quoted in *Uniscience Daily,* "Frog Venom Could Prove Vital in War Against Disease," September 20, 2001. www.frogs.org.

41. Bob Johnson, "Frogs and Educators Helping Toads," paper issued by the Toronto Zoo, Canada, 2002.

42. Sam Lee, senior keeper at the Wildlife Conservation Society, Department of Herpetology, Bronx, NY, phone conversation with author, September 13, 2002.

43. Kevin Buley, herpetology department head, Durrell Wildlife Conservation Trust, Jersey Zoo, British Channel Islands, e-mail correspondence, August 5, 2002.

44. Kevin Zippel, curator of National Amphibian Conservation Center, Detroit Zoo, Michigan, phone conversation with the author, September 13, 2002.

45. Lips, phone conversation with author.

46. Lips, phone conversation with author.

Glossary

---

acid rain: Rain containing acids that form in the atmosphere when industrial gas emissions combine with water.

amplexus: The act in which a male frog holds onto a female, waiting to fertilize her eggs.

anurans: Members of the order Anura, one of three orders of amphibians, considered the tailless amphibians. All frogs and toads belong to this order.

biomonitor: A plant or animal that is easily studied and quickly shows signs of distress when a habitat is compromised.

Bufonidae: The family of frogs that is considered "true toads."

chromatophore: A pigment-containing or pigment-producing cell that by expansion or contraction can change the color of the skin.

chytrid: A fungus common to the environment. One particular species of chytrid is considered by scientists to be responsible for the decline of many species of frogs and toads.

CITES: Convention on International Trade in Endangered Species (of wild fauna and flora). CITES is responsible for regulating trade in species that are considered in danger of extinction.

dermis: The deep sensitive layer of the skin beneath the epidermis.

desert: A dry, often sandy region of little rainfall, extreme temperatures, and sparse vegetation.

dissection: The act of cutting apart or separating for the purpose of anatomical study.

diurnal: To be most active during the daylight hours.

El Niño: A warming of the ocean surface off the western coast of South America that occurs every four to twelve years, creating unusual weather patterns in various parts of the world.

endemic: A species that is confined to a particular region and found nowhere else in the world.

epidermis: The outer, protective, nonvascular layer of the skin of vertebrates, covering the dermis.

global warming: An increase in the average temperature of Earth's atmosphere, especially a sustained increase sufficient to cause climatic change.

grasslands: Ecosystem in which grasses are the dominant vegetation. Temperate grasslands have a rainfall between ten and thirty inches, a high rate of evaporation, and seasonal and annual droughts. Tropical grasslands develop in regions with marked wet and dry seasons.

herpetologist: A scientist that studies reptiles and/or amphibians.

keratin: The protein that makes up human hair and nails as well as parts of amphibian anatomy.

metamorphosis: The transformation of a frog's body from the larval form of tadpole to the adult stage of frog.

mucous glands: The glands on a frog's skin that secrete a slimy substance that keeps the frog's skin moist.

nuptial pads: Pads that form on the feet of some species of male frog in order to allow the frog to hold on to the female during amplexus.

ozone: The layer of the atmosphere that blocks ultraviolet radiation.

paratoid glands: Glands that secrete a distasteful and toxic fluid mainly in species of the genera bufo.

rainforest: A dense evergreen forest with constant warm temperatures and an annual rainfall of six to thirty feet with no dry season.

Ranidae: The family of frogs that is considered "true frogs."

94

respiration: The act or process of exchanging gases with the environment in order to consume oxygen and release carbon dioxide.

spawn: To lay eggs in great numbers.

SSP: Species Survival Program, a program designed by the American Zoo and Aquarium Association to oversee the breeding and conservation efforts of a particular species.

terrestrial: To live on the ground.

trematodes: A type of parasitic flatworm.

ultraviolet radiation: High energy radiation from the sun that has the capability to damage organisms.

vertebrates: Animals with a backbone.

wetland: A lowland area, such as a marsh or a swamp, that is saturated with moisture.

Organizations
to Contact

American Zoo and Aquarium Association
8403 Colesville Road
Suite 710
Silver Spring, MD 20910-3314
(301) 562-0777
website: www.aza.org

American Zoo and Aquarium Association is a nonprofit organization dedicated to the advancement of zoos and aquariums in the areas of conservation, education, science, and recreation.

Amphibian Conservation Alliance
c/o Ashoka
1700 North Moore Street, 20th Floor
Arlington, VA 22209
(703) 807-5588
e-mail: info@frogs.org
website: www.frogs.org

Amphibian Conservation Alliance is a partnership between scientists and policy experts whose mission is to protect frogs and other amphibians worldwide.

Environment Australia
John Gorton Building
King Edward Terrace
Parkes ACT 2600
GPO Box 787
Canberra ACT 2601
61 2 6274 1111
website: www.ea.gov.au

Environment Australia advises the commonwealth government on policies and programs for the protection and conservation of the environment.

Frogwatch USA
Coordinator: Amy Goodstine
National Wildlife Federation
1400 16th Street NW, Suite 501
Washington, DC 20036
(202) 797-6891
e-mail: frogwatch@nwf.org
website: http://monitoring2.er.usgs.gov

Frogwatch USA is an educational, long-term frog and toad monitoring program coordinated by the National Wildlife Federation and the U.S. Geological Survey's Patuxent Wildlife Research Center.

U.S. Fish and Wildlife Service
Department of the Interior
1849 C Street, Room 3012
Washington, DC 20240
(202) 208-5634
website: www.fws.gov

The service is the lead federal agency in the conservation of threatened and endangered species. It operates programs to educate Americans on fish and wildlife resources as well as to assist state governments in the conservation of wildlife.

For Further Reading

Books

Trevor Beebee, *Frogs and Toads.* London: Whittet Books, 1985. An excellent overview of the natural history of anurans with emphasis on English anurans.

Harold Cogger and Richard G. Zweifel, *Reptiles and Amphibians.* New York: Smithmark, 1992. This book covers both amphibians and reptiles with a nice section on amphibians and excellent photographs.

Kim Long, *Frogs: A Wildlife Handbook.* Boulder, CO: Johnson Books, 1999. A nice combination of field guide, fact book, and folklore collection.

Rob Nagel, *Endangered Species.* Vol. 3. Detroit: UXL, 1999. This volume of the *Endangered Species* encyclopedia set has a good overview of several endangered frogs as well as an opening discussion on classification and what makes a species endangered.

Websites

AmphibiaWeb (http://elib.cs.berkeley.edu). This site, inspired by global amphibian declines, allows free access to information on amphibian biology and conservation.

Declining Amphibian Population Task Force (DAPTF) (www.open.ac.uk). The DAPTF was established in 1991 by the Species Survival Commission (SSC) of the World Conservation Union (IUCN). This website contains their past newsletters and recent press releases.

Frogland (http://allaboutfrogs.org). A private site that has good information as well as a lot of fun frog trivia, games, and other entertainment.

Tadpole on the Hop (http://abc.net). A great page on tadpole transformation, an overview of Australian frog decline and discussion of the famous green and golden bell frog.

A Thousand Friends of Frogs (http://cgee.hamline.edu). A great website created by the Center for Global Environmental Education at Hamline University Graduate School of Education. This site has many resources for both teachers and students.

Toads Dome (www-astro.physics.ox.ac.uk). This is a site that is dedicated to toad information both for science and for fun.

Works Consulted

Books

T.J.C. Beebee, *Ecology and Conservation of Amphibians.* London: Chapman and Hall, 1996. An overview of amphibian natural history including the current pressures on populations. Also addresses conservation efforts.

Marty Crump, *In Search of the Golden Frog.* Chicago: University of Chicago Press, 2000. The story of the disappearance of the Costa Rican golden frog told by the researcher who discovered its loss.

William E. Duellman et al., *Patterns of Distribution of Amphibians: A Global Perspective.* Baltimore: Johns Hopkins University Press, 1999. A series of scientific papers detailing the distribution of amphibians worldwide.

William E. Duellman and Linda Trueb, *Biology of Amphibians.* Baltimore: Johns Hopkins University Press, 1994. This text is considered the standard textbook for college amphibian study.

W. Ronald Heyer et al., *Measuring and Monitoring Biological Diversity, Standard Methods for Amphibians.* Washington, DC: Smithsonian Institute Press, 1994. A manual designed to standardize field studies involving amphibians.

Robert Hofrichter, *Amphibians.* Buffalo: Firefly Books, 2000. A fantastic and up-to-date book that covers amphibian natural history as well as challenges to amphibian survival.

Chris Mattison, *Frogs and Toads of the World.* Dorset, UK: Blandford Press, 1987. A nice overview of frogs and toads from around the world.

Roy W. McDiarmid et al., *Tadpoles, the Biology of Anuran Larvae.* Chicago: University of Chicago Press, 1999. An extensive study of the behavior, biology, and anatomy of tadpoles.

Kathryn Phillips, *Tracking the Vanishing Frogs.* New York: St. Martin's Press, 1994. A personalized overview of the problems and mysteries surrounding the disappearance of many species of frogs and toads.

William Sounder, *A Plague of Frogs.* New York: Hyperion, 2000. A personal narrative covering the stories surrounding mutated and disappearing frogs and toads.

Michael J. Tyler, *Australian Frogs and Toads, a Natural History.* Ithaca, NY: Cornell University Press, 1994. An excellent book covering Australian anurans, their uniqueness and ecology.

Periodicals

Chris Amodeo, "Frogs Under Fire," *Geographical,* vol. 74, no. 4, 2002.

Jessica Burton, "How to Make a Breeding Pond for Your Toads and Frogs!" *Organic Gardening,* vol. 41, no. 5, 1994.

Ron Cowen, "Brooding over Australian Frogs," *Science News,* March 3, 1990.

Tina Hesman, "Children's Restoration Project Brings a Bit of Wetland to Universal City," *St. Louis Post,* May 27, 2002.

Bob Johnson, "Frogs and Educators Helping Toads," paper issued by the Toronto Zoo, Canada, 2002.

Bob Johnson, *Recovery of the Puerto Rican Crested Toad, "Peltophryne lemur:* Ex Situ and In Situ Population Management," Toronto Zoo, Canada, 2002.

Suzanne B. Kaplan, "Let's Hear It for the Creepy Crawlies," *Discover,* September 2000.

M.J. Lannoo et al., "What Amphibian Malformations Tell Us About Causes," *American Zoologist,* vol. 40, no. 6, 2000.

Kathrin Day Lassila, "The New Suburbanites: How American Plants and Animals Are Threatened by Urban Sprawl," *Amicus Journal,* vol. 21, no. 2, 1999.

Sam Lee, "The Kihansi Spray Toad: Program Overview and Captive Propagation Guidelines," paper prepared for the Wildlife Conservation Society, January 15, 2002.

Sam Lee, "The Kihansi Spray Toad Report No. 1," quarterly report of the Wildlife Conservation Society, September 19, 2001.

Sam Lee, "The Kihansi Spray Toad Report No. 4," quarterly report of the Wildlife Conservation Society, July 15, 2002.

Beth Livermore, "Amphibian Alarm: Where Have All the Frogs Gone?" *Smithsonian,* vol. 23, no. 7, 1992.

Jon R. Luoma, "Disappearing Frogs and Toads," *Audubon,* May/June 1997.

Ashley Mattoon, "Amphibia Fading," *World Watch,* vol. 13, no. 3, 2000.

Judith Maunders, "Silent Streams: Although We Now Know Why Frogs Are Dying, Their Future Is Far from Assured," *Ecos,* October–December 2001.

Susan Milius, "Fatal Skin Fungus Found in U.S. Frogs," *Science News,* July 4, 1998.

Susan Milius, "Frog Real Estate: More than Location," *Science News,* January 3, 1998.

Susan Milius, "Wafting Pesticides Taint Far-Flung Frogs," *Science News,* December 16, 2000.

Rachel Nowak, "Leave Well Alone," *New Scientist,* September 2, 2000.

Per Nystrom et al., "The Influence of Multiple Introduced Predators on a Littoral Pond Community," *Ecology,* vol. 82, no. 4, 2001.

Stephanie Pain, "Ready to Croak: A Fungal Disease Is Wiping out New Zealand's Ancient Frogs," *New Scientist,* April 6, 2002.

Matthew J. Parris, "High Larval Performance of Leopard Frog Hybrids: Effects of Environment-Dependent Selection," *Ecology,* vol. 82, no. 11, 2001.

Pamela Parseghian, "Frogs' Legs: The Prime 'Rib-it' of the Amphibian World," *Nation's Restaurant News,* February 14, 2000.

Kathryn Phillips, "Sun Blasted Frogs," *Discover,* vol. 16, no. 1, 1995.

Kathryn Phillips, "What's Wrong with the Frogs?" *Sierra,* March 1999.

Joanna Poncavage, "Let Toads Tackle Your Insect Pests!" *Organic Gardening,* May/June 1994.

Sahotra Sarkar, "Ecology Theory and Anuran Declines," *BioScience,* vol. 46, no. 3, 1996.

Jason B. Searle, "Hey, Who Turned the Water Off?" *Wildlife Conservation,* vol. 104, no. 1, 2001.

U.S. Fish and Wildlife Service and American Zoo Association Crested Toad SSP, "Puerto Rican Crested Toad Conservation Meeting Minutes," December 4, 2001.

S. Weisburd, "Jump for Joy: Blue Frog Babies," *Science News,* April 16, 1988.

Tammie Wersinger, "For Rare Toad to Survive, Breeding Is Fundamental," *Orlando Sentinel Tribune,* August 1, 2002.

Internet Sources

Jeff Barnard, "Frogs, Toads and Salamanders Rapidly Declining Since 1960s," *Associated Press, ABCNews.com,* 2000. http://more.abcnews.go.com.

Chicago State University, "The Somewhat Amusing World of Frogs." www.csu.edu.

Stentor Danielson, *National Geographic News,* July 9, 2002. http://news.nationalgeographic.com.

Beverly Billings Fenn, "It's Raining Frogs and Toads," *Colorado Herpetology Society,* 1998. www.coloherp.org.

Eryn Gable, "EPA Asked to Ban Atrazine, Investigate Manufacturer," *Greenwire,* June 4, 2002. www.frogs.org.

Ben Ikenson, "A Harbor in the Desert," *Environmental News Network,* November 28, 2001. www.frogs.org.

Mary Losure, "MPCA Backs Away from Frog Research," *Minnesota Public Radio,* June 12, 2001. http://news.mpr.org.

Mary Losure, "Vanishing Frogs of the Panamanian Rain Forest," *Minnesota National Public Radio,* June 2, 1998. http://news.mpr.org.

Kim Y. Masibay, "Rainforest Frogs: Vanishing Act? Frog Populations Around the World Are Dying Off Mysteriously. Can Scientists Save Them?" *Science World,* March 11, 2002. www.findarticles.com.

Alexandra Ravinet, "Listening Hard for Frogs," *ABCNews.com,* July 2, 1999. http://more.abcnews.go.com.

Jack Rosenberger, "Harvest of Shame: Dissection's Deadly Toll Hits Frogs the Hardest." *E: The Environmental Magazine,* July/August 1998. www.findarticles.com.

Michon Scott, "Where Frogs Live," *Goddard Space Flight Center,* March 5, 2001. http://earthobservatory.nasa.gov.

Susan Tweit, "Spadefoots and Termites," *Southern New Mexico Online.* www.southernnewmexico.com.

Uniscience Daily, "Frog Venom Could Prove Vital in War Against Disease," September 20, 2001. www.frogs.org.

U.S. Fish and Wildlife Service, "Chirichua Leopard Frog Listed as a Threatened Species," June 13, 2002. http://news.fws.gov.

Wendy Williams, "Flash and Thunder (Frogs of Madagascar)," *Animals,* July 2000. www.findarticles.com.

Websites

Exploratorium (www.exploratorium.edu). Great information on frogs based on a temporary exhibit at the Exploratorium, an online museum of science and art sponsored by the renowned San Francisco Exploratorium.

Project Golden Frog (www.projectgoldenfrog.org). This website gives an in-depth look at the problems confronting the golden frog as well as the efforts to preserve this species.

Index

Picture Credits

About the Author

Rebecca K. O'Connor is a professional animal trainer with a degree in creative writing and a fascination for all living creatures. Currently she works at a desert zoo in Southern California. She developed a passion for anurans and an interest in amphibian decline while "frogging" with friends in Australia. She has been hopping after frogs and toads ever since.